IMPRONTA

The Cognitive Heuristics of Nature

IMPRONTA

The Cognitive Heuristics of Nature

JAVIER VALLE

ISBN: 9798297010338
INDEPENDENTLY PUBLISHED

FOR MY SON

TABLE OF CONTENTS

FOREWORD

Back in my childhood, we used to visit my Uncle Mac, my dad's older brother, at his farm. During bean harvest season, he'd rig up these wooden and mesh sifters—simple yet ingenious contraptions that helped us separate what was useful from what was not. The first stage required a coarse, rugged sifter. We would pile the dry pods onto it and strike them with a rhythmic, repetitive beat—a forceful act that shattered their shells. It was the Big Bang in miniature: an explosion of movement, impact, and resonance that set everything else in motion. And as more of us joined in, each sifter becoming its own small drum, a strange thing would happen—we'd fall into rhythm. Without trying, we'd sync up as if guided by some invisible metronome. Even if someone stepped away and came back, they'd fall right back into the collective beat. The beans would fall through the wide mesh into a square bin beneath, while the empty husks stayed behind.

Then came the refining step. We'd pour the beans into a second, more delicate sifter and shake it from side to side. The small stones, dirt, bits of stem, and dry skin would sift through the mesh. Left on top were the clean, usable beans—a universe of clean beans.

That sifter, in all its simplicity, was a miniature model of the cosmos. Much later, I realized that everything that exists has passed through a similar kind of filter—a cosmic sieve that selects, discards, and shapes reality at every moment.

The first great sieve operated at the dawn of existence. There was no matter yet, only pure energy—fluctuations in quantum fields, vibrations rippling through the raw fabric of newborn space-time. Most of those vibrations faded into nothing. But some aligned just right, resonated with one another, and amplified each other enough to persist. From that coherence, particles emerged.

The Higgs field acted as the mesh of that primordial sieve. Particles that "tangled" in its net acquired mass—like the good beans that stayed on top. Others, like photons, slipped through effortlessly, like fine dust. The sieve didn't create the bean, but it sealed its fate by granting it the property that defined its journey: its weight, its mass.

Then came the chemical sieve. Particles formed atoms, and atoms joined into molecules. With each step, the laws of physics filtered out the unstable combinations. What survived were the arrangements that held together in the face of surrounding forces. That persistence is a kind of proto-memory: a trace encoded in structure itself. A vibration becomes a form, and that form constrains what can happen next. No intent, no awareness—just physics choosing, by way of resonance, what's allowed to continue.

This process gave way to an even finer sieve: the sieve of information. The natural selection of stable configurations can, in fact, be measured. Biologist Jack Szostak proposed a way to calculate what he called "functional information"—how rare and specific a structure is that fulfills a purpose, compared to all possible alternatives [Szostak2012].

Applied to Earth's minerals, the results are staggering. Out of a practically infinite number of atomic combinations—a 1 followed by 46 zeros—only about 6,000 configurations have managed to form stable minerals we actually find today. The information packed into that selection is about 142 bits. It may not sound like much, but with just two bits—zero and one—we built the entire digital age. Nature, with its 142 bits of mineral logic, built the geology of a planet.

Life itself is the most complex sieve we know. It doesn't just adapt to existing niches; it invents them, massively expanding the range of possible combinations and accelerating the accumulation of functional information. To say that evolution began with the cell is to overlook the deeper stages, where relentless selection was already at work—not of organisms, but of patterns.

Which brings us to the final sieve: us. For the first time, one of the universe's creations has learned how to build its own filters. We're changing the way functional information accumulates. Consider the work of Juan Pérez-Mercader, a researcher at Harvard's Origins of Life Initiative, who succeeded in forming primitive cellular structures using nothing but chemicals and LED light [PérezMercader2021].

We're not just on the verge of creating life—we're designing new sifters, new fields of selection. More than any discovery before, this makes us an evolutionary force, with the power—and the responsibility—to decide what gets filtered through from now on.

INTRODUCTION

Imagine a hummingbird. Before dawn, its body is already operating with an efficiency molded over millennia, "calculating" with stunning precision. It knows where and when to find the most abundant nectar, using evolutionary shortcuts to make quick, effective decisions. This hummingbird is not an exception. Life, from bacteria to human beings, doesn't always rely on exhaustive analysis. Success is based on adaptive heuristics—simple, efficient rules, forged by millions of years of natural selection, that guide behavior even in the absence of a brain—which represent life's practical intelligence, encoded in biology itself, and manifested in the complex information-processing systems we know as brains.

Adaptive heuristics are behavioral patterns or shortcuts that have emerged and been maintained throughout the evolution of species. Their value lies in allowing organisms to process information and make decisions efficiently in complex, dynamic environments where time and resources are limited, minimizing energy expenditure for the sake of survival and reproduction. They are the practical intelligence of life, encoded in biology itself and not just in neuroanatomy.

Consider the *Mimosa Pudica*, that plant that retracts its leaves at the slightest touch. This act, an apparent mechanical "reaction," is a defensive heuristic: a rapid, energetic response to what the environment perceives as a threat, minimizing the risk of physical damage or predation. There is no mind in the human sense, but there is a biological "decision" for energy conservation and survival.

Or consider a colony of ants that, when foraging, doesn't map out the entire terrain but follows a simple rule: "follow the pheromone trail." This "follow the leader" heuristic leads to astonishing collective efficiency, allowing the colony to optimize resource collection without the need for centralized planning.

Even further, consider a virus. This entity, lacking cells or its own metabolism, "appropriates" a host's machinery to replicate itself with brutal efficiency. This functional manipulation, far from being an accident, is an adaptive heuristic of energy minimization in its purest form: the virus externalizes almost all of its biological "work," optimizing its survival and replication at minimal self-cost. It is a master of biological economics.

These examples reveal that the principles of adaptation are truly universal. Adaptive heuristics are the fundamental "software" that allows living systems, at all levels of complexity, to function effectively. They are the functional principles that govern key dynamics such as:

- A system that regulates motivation and reward: activating neurochemical salience modules in the face of relevant events, a common mechanism in animals, but also observable in bacteria through chemotaxis systems.

- Attention is focused as part of a cognitive economy strategy, highlighting stimuli of immediate adaptive value. This principle governs everything from tropistic responses in plants to visual focus in predators.

- Prior knowledge is reused and restructured as part of learning by analogy and hierarchical encoding. This is a highly conserved heuristic in animal brains and in distributed networks, such as swarms.

- The organization of information occurs at multiple levels: from spatial maps in insects to the clustering of chemical signals in cells, facilitating faster access and more reliable decisions.

- Adaptive failure and play constitute controlled environments for exploration. They are energetically measured trials that allow for flexibility without compromising the system's integrity, as demonstrated by synaptic plasticity in mammals and behavioral adjustment in simple organisms.

These manifestations can be grouped into major functional families of adaptive heuristics, which will be developed in more detail in later essays:

v

- Fundamental processing and survival heuristics.
- Learning and cognition heuristics
- Bias and judgment heuristics
- Social and interaction heuristics
- Resilience and advanced adaptation heuristics.

Each of them can be observed in organisms without a nervous system, such as in mycelial networks, as well as in highly complex human cognitive systems.

Understanding these adaptive heuristics offers us a new lens for observing life. It invites us to recognize that intelligence and adaptations are not just properties of complex brains, but an omnipresent evolutionary legacy. In the following essays, we will delve into how these rules of life manifest and intertwine in the complex dance of social interaction, collective intelligence, and creativity, both in nature and in our own lives, revealing the profound wisdom that underlies every living being.

Universal cognitive system

If adaptive heuristics are the software of life, then we must ask: what is the operating system that runs them? What connects bacteria, fungi, animals, plants, and humans is not a common anatomical structure, but a functional architecture: a system capable of capturing information from the environment, processing it, generating a response, and receiving feedback on its effects. We can call this functional and decentralized system the Universal cognitive system.

The Universal cognitive system is not limited to neuronal activity. It includes any network of biological processing that, through sensory, chemical, or structural circuits, performs operations we could recognize as cognitive: making decisions, anticipating events, modulating responses, forming memories or patterns.

An immune system that "remembers" a virus. A plant that changes its morphology after several attacks. A fungus that regulates the growth of a tree's roots. A colony of insects that distributes tasks efficiently. All of them participate in this system, and all implement adaptive heuristics.

This model allows us to redefine cognition as a distributed, embodied, and evolutionary phenomenon, not exclusive to the human brain. The Universal cognitive system operates as a decentralized Bayesian network, where different biological "agents" process local information, share signals, and produce a global response coherent with the environment. This network is motivated by a principle shared throughout nature: the minimization of unnecessary effort.

In this sense, the principle of least action (visible in both classical and quantum mechanics) becomes an evolutionary metarule: living systems evolve to find paths of action that reduce energy expenditure, increase the predictability of the environment, and optimize their adaptive performance. Predictive error and surprise are not failures, but engines of learning.

This approach also highlights the idea that the cognitive "agents" in the human brain (such as Theory of Mind, the salience system, or the default network) function as competing and consensual modules within a distributed predictive architecture. It is a model of co-management of perception, action, and learning, which has echoes at all levels of life.

Therefore, when we talk about adaptive heuristics, we are no longer just talking about neuroscience or animal evolution, but about a biological epistemology: a way of knowing and acting in the world that has emerged repeatedly because it is functional, elegant, and resilient. In the pages that follow, we will explore how this system manifests, and what lessons we can extract for our own lives, organizations, and shared futures.

PART I

THE INNER ARCHITECT

Origins

INVISIBLE WORLD

In the tapestry of nature, life buzzes in a symphony of interactions, many of which escape our superficial perception. From the buzz of a bee to the stealthy stalk of a predator, each creature reveals an ingenuity that, if we learn to observe it, uncovers profound truths about the universal principles of intelligence and adaptation.

The meadow, with its fragrance and bustle, is a testament to this complexity. Flowers, in their quest for perpetuation, attract bees with the promise of nectar, in a vital exchange. However, in this delicate balance, the orchid emerges as a master of deception. With its striking color and yellow stamen, it tempts bees without offering a reward. A thirsty bee, upon inserting its head into the flower, encounters a firmly attached stamen. Although it continues its flight without nectar, it transports the pollen to another flower. The orchid has achieved its purpose.

Here there is no exchange, but a brilliant adaptive heuristic: an efficient rule, forged by evolution, that optimizes the result with astonishing resource economy. It is, in essence, one of nature's most elegant Minimum Viable Products (MVPs); the flower presents the minimum necessary stimulus to activate the bee's behavioral program. This act of cunning dismantles our assumptions and forces us to ask a fundamental question: how do these principles of efficiency and adaptation operate in even more intricate systems, such as the brain, where information is constantly processed and transformed to build experience?

This silent and efficient functional intelligence is not an anomaly. The potter wasp, for example, builds its exquisite nest and provisions it with paralyzed caterpillars, not through individual learning, but through prior knowledge inscribed in its lineage. Its action is a form of anticipatory simulation: a strategy in the present to ensure a future it will never see. In the stalk of the weasel or the flight of the mouse, we see two organisms operating as prediction machines, where each movement is an attempt to minimize fatal error.

Perhaps the most radical example is offered by the bat. It navigates in total darkness through echolocation, building a reality from echoes that, to us, are silence. Its experience demonstrates conclusively that there is no single objective reality, but that the brain is the editor of each species' reality, a constructor of worlds tailored to their adaptive needs.

What connects the deceiving orchid, the building wasp, and the navigating bat is not anatomy, but a common operational logic. They are manifestations of the same evolutionary engine: an Inductive Machine, a universal system that life has used since its inception to learn and adapt through the resolution of contradictions. In the face of an error, an ambiguity, or a disturbance, this engine activates a search for new solutions, and those that are successful are consolidated as new rules of behavior.

This is the endless game of adaptation.

This book is a journey to uncover the rules of that machine. Throughout these pages, we will explore how these adaptive heuristics operate at all levels of life, from cells to societies. We will not look for intelligence where we have always believed we would find it, but will discover it as an emergent property of life itself, a symphony of patterns that organize and persist.

This is an invitation to acquire a new lens to observe the world, to recognize the functional code of the Architect Within who, silently, shapes every living being and, ultimately, the architecture of our own mind..

THE ENDLESS GAME

Peter Putnam, a 20th-century physicist and philosopher, developed a unique mathematical model known as the "inductive machine." This model differs from the Turing machine in that it is capable of generating its own internal rules, adapting to its environment, and establishing its own goals through continuous interaction. Putnam's proposal offers a unifying framework for understanding intelligence and adaptation across a wide range of biological scales, from bacteria and plants to the future of human cognition augmented by artificial intelligence. His model provides a coherent lens through which we can understand how living systems—and, ultimately, brain-AI hybrids—formulate and perfect their own internal rules, adapting to a constantly evolving universe.

Its principles are fundamental: the system is composed of a vast number of binary units that operate in parallel, switching on and off. In an organism, these units can be neurons; at a deeper level, molecular components. These units do not act in isolation; they are interconnected by conditioned reflexes, where the action of one influences the probability of another's activation. When a chain of these units forms a self-reinforcing loop, this sequence of activation becomes a "movement" or a "rule" emitted by the system.

The engine of this inductive machine is infinite repetition or existence. This is not a programmed goal, but one inherent to the very nature of being: to persist from one moment to the next is to repeat one's own existence. Any deviation from this state of continuity is a "perturbation" from the environment. The "game" of induction consists of the system, in the face of a perturbation, having to emit the correct sequence of "movements"—the self-reinforcing loops—to alter the environment in such a way that it returns to its original state.

In this game, learning, or induction, emerges from experience. Initially, the system's responses may be random. However, when a particular "movement" succeeds in silencing a perturbation (i.e., the system returns to its state of repetition), that successful loop is "connected" or "recorded" in the system as a behavioral "rule." The subsequent application of this rule becomes deductive, allowing the system to react more efficiently.

Let's observe a fruit fly larva when it first emerges from its environment. Imagine that, after growing inside a mango, its movements expel it and it falls to the ground. At first, its tiny body, similar to a wrinkled grain of rice, explores the environment with hesitant movements of its "head," alternating turns to the left and right. With this movement, the larva chooses its route based on the level of humidity, an inherited instinct from its first moments of life inside the mango. If it encounters an excessively humid area—which from its perspective is a "pit" or a "pond"—this "perturbation" hinders its progress and threatens its "state of repetition" (to move effectively). It then redefines its strategy and opts for the path with a lower concentration of water. As it progresses, its system "learns" to distinguish between dry, damp, and flooded, adjusting its "movement" to minimize resistance and resume its continuous advance. This seemingly trivial micro-event encapsulates the essence of the inductive machine: an adaptive system immersed in a "perpetual interaction" with its environment.

But the real magic happens when a contradiction arises: two previously successful rules are invoked in parallel, competing to generate mutually exclusive movements. Instead of collapsing, the system, driven by the perturbation the contradiction generates, is forced to look for new variables or contextual information. This leads it to "reshape" the original loops and forge a new rule, more general and nuanced, that resolves the inconsistency. It is through this resolution of contradictions that the system not only learns, but innovates and ascends the "ladder of increasingly intelligent behavior."

Putnam emphasizes a crucial point: this is a "game that is never won." The system never achieves a state of perfect repetition because the environment is in constant flux and perturbations are inevitable. However, it is precisely this impossibility of "winning" that drives continuous adaptation. By incessantly trying to repeat its state, the system transforms, adapts, and, ultimately, innovates. This idea—that existence is a perpetual struggle and that fulfillment does not reside in the stillness of a final solution, but in the dynamism of the search—resonates throughout the centuries in various philosophical, spiritual, and existential discourses, from Eastern philosophies of impermanence to Western notions of progress. It is a reflection of the condition of life itself: an incessant process of being and becoming.

The genius of the inductive machine lies in its ability to transcend scale and offer a unifying logic for intelligence and adaptation in diverse biological systems, from the microscopic to the macroscopic. At its most fundamental level, the individual neuron embodies this logic. Bombarded by thousands of synaptic inputs that act as constant "perturbations," the neuron processes this information and, through complex integration mechanisms, decides whether or not to "fire." The connection patterns that lead to successful outcomes (such as the effective transmission of a relevant signal) are reinforced by synaptic plasticity, recording communication "rules." When contradictory signals arrive simultaneously, the neuron's ability to weigh them and generate a coherent response is a micro-resolution of contradictions that creates finer connection "rules."

Scaling up in complexity, a bacterium can be seen as an "inductive machine with multiple nuclei." Although it lacks neurons, its intricate molecular and genetic machinery operates under the same principles. Each protein that changes state or each gene that is activated is a "binary unit." The biochemical pathways are the "self-reinforcing loops," and its goal is survival and replication. In the face of an environmental perturbation, such as nutrient scarcity, the bacterium "makes movements"—metabolic changes, gene expression—to return to its state of existence. Bacterial

5

Quorum Sensing is a fascinating manifestation of this: through the detection of signaling molecules (evidence), bacteria "infer" population density and update their "belief" about the right time for collective behavior, such as forming a biofilm [Bassler2002]. It is a primitive Bayesian consensus, where multiple "nuclei" contribute to a collective decision.

Plants, without a central nervous system, extend this logic to a network of interconnected cells. Each cell or cell group perceives "perturbations" from the environment (light, humidity, pathogens) and "acts" by adjusting its growth and physiology. The coordination of these responses through hormones and electrical signals is a type of distributed cellular Quorum Sensing, where the plant collectively "inferrs" the best strategies to optimize its survival. It is a massively parallel Bayesian system, adapting to its ecological niche [Hebb1949].

Octopuses offer an even more sophisticated example of distributed intelligence. Although they have a central brain, much of their motor and cognitive control resides in their eight arms, each with semi-independent autonomy. Each arm can be considered a local "network of nuclei" of the inductive machine, performing Bayesian inferences about the immediate environment (texture, shape). The central brain, for its part, integrates these local inferences to make global decisions, acting as an orchestrator that seeks "consensus" among the "beliefs" of its multiple appendages, which allows for astonishing flexibility and adaptability [GodfreySmith2016].

Finally, the virus, despite its simplicity, fits into this logic as a "parasitic inductive machine." It does not learn at the individual level, but its genome is the "rule" resulting from an evolutionary "endless game." Mutations are its "random variations," and replication in a host is its "self-reinforcing loop." The host's defenses are the "perturbations" that force the virus to "resolve contradictions" through the natural selection of new "rules" (successful mutations) to maintain its existence [Rolls2010]. The virus is the ultimate expression of a strategy of specialization and efficiency.

If the neuron is an inductive machine and a bacterium is a multinuclear machine, then the human organism is not only a galaxy of interconnected machines, but an intrinsic multi-agent system that operates through layers of unprecedented complexity, embodying the known pinnacle of a Bayesian system. This architecture, increasingly recognized by contemporary neuroscience, allows the organism not only to react to the world, but to model, predict, and adapt in extraordinary ways.

The human organism is a master of Bayesian inference, constantly updating its "beliefs" and predictive models of the world based on an incessant flow of new evidence. Composed of networks of "multinuclear machines" (functional modules), it processes multimodal information (sensory, emotional, contextual, abstract) to form sophisticated judgments about the probability of future events and the validity of different interpretations of reality. The "quorum" in the human organism is not a simple binary signal, but a probabilistic convergence of billions of neurons and networks that compete and collaborate to arrive at a conscious decision [Singer1999]. The resolution of contradictions, as proposed by the Inductive Machine, manifests in the human ability to weigh moral dilemmas, solve complex scientific problems, or reconcile opposing beliefs, leading to the formation of new understandings and broader conceptual frameworks.

This capacity extends to the evolution of our most basic heuristics. The "Fight or flight" responses (or the expanded "FFF+": Fight, Flight, Freeze, Fawn) are primary heuristics, burned into our lineage by millions of years of evolutionary play. However, the increasing complexity of social life and the development of accumulated knowledge have spurred the emergence of higher-order adaptive heuristics: self-regulation. This ability, which allows us to modulate impulses, manage emotions, plan for the long term, and learn from experience consciously, is a complex "rule" forged in the uninterrupted game of adapting to an increasingly

7

THE ENDLESS GAME | **PART I**

multifaceted environment. It is the result of countless "games" where social and cognitive "perturbations" have forced the system to find new variables and create more general "rules" to maintain coherence and advance.

This is where artificial intelligence comes into play as the next evolutionary frontier. The vision of our organism, augmented by devices that combine our biological reactions with AI to achieve a "Quorum Sensing" with other AIs, represents a monumental evolutionary thrust. AIs would not make decisions for us, but would act as consensus agents, drastically expanding our ability to process "evidence" and resolve "contradictions." This hybrid "Quorum Sensing" would allow an unprecedented acceleration of the induction "game." The AI agents could:

- Process data at scales and speeds unimaginable for the biological organism, presenting "perturbations" and "variables" from a vast information environment.

- Identify patterns, correlations, and biases that would escape our human perception, enriching the database for the Bayesian update of our beliefs.

- Simulate multiple scenarios and probable outcomes, helping our organism to mentally "test" various "rules" or solutions before action.

- Enhance self-regulation, offering real-time feedback on our cognitive and emotional states, and suggesting strategies to optimize mental performance.

By integrating these capabilities, we will not only increase the speed of our decision-making, but we will also forge more sophisticated and robust heuristics, capable of navigating dilemmas of unprecedented complexity, from climate change to the frontiers of bioethics. It would be a co-evolution, a synergy where AI helps us play the game of adaptation more efficiently.

The Inductive Machine offers us an extraordinarily powerful lens for understanding intelligence and adaptation. From the exploratory turns of a worm searching for moisture to the intricate decision-making of an octopus, the resilience of a plant, the molecular survival of a bacterium, or complex human cognition, we see the same fundamental principles in action: a system that, immersed in an "endless game" with its environment, adapts, learns, and generates order through variation, selection, and the resolution of contradictions. The "game that is never won" is not a failure, but the driving force of innovation and evolution.

In this context, the emergence of artificial intelligence should not be seen as a competitor or a replacement for human intelligence, but as a natural extension of this evolutionary process. Just as self-regulation emerged as a higher-order heuristic to manage social complexity, the symbiosis between the human organism and AI promises to take the "game of induction" to an unprecedented level. The vision of a "Quorum Sensing" between our Bayesian organism and AI agents suggests a future where decision-making becomes exponentially faster, more informed, and more adaptable. AI, acting as a vast field of "sensors" and "analyzers" of information, can accelerate the cycle of variation and selection of heuristics in our own thinking, helping us forge more refined "rules" to interact with an increasingly complex world.

This is not simply an increase in processing capacity; it is an evolutionary push that could redefine what it means to learn, adapt, and innovate. Intelligence, viewed through this lens, is a continuous process, an incessant adaptation to the perturbations of existence. In this fascinating journey, from humble cellular beginnings to the conscious mind, the next frontier is not the creation of an autonomous AI that replaces us, but the co-evolution towards a form of cognition that integrates the best of biological and artificial intelligence, allowing us to play the "endless game" of life with unprecedented wisdom and resilience.

THEORY OF MIND

Who Thinks of the Other?

Theory of Mind (ToM) is an intrinsic and crucial cognitive capacity that allows individuals to infer and attribute complex mental states—including beliefs, desires intentions, emotions, and knowledge— not only to themselves but also to others. This ability involves recognizing that the mental states of other people can be different from and not necessarily aligned with one's own, underscoring a profound understanding of individual subjectivity.

This faculty is, without a doubt, a cornerstone for the development of meaningful and adaptive social interactions. Without ToM, communication, empathy, cooperation, and conflict resolution would be extremely limited or even impossible processes. It allows us to predict the behavior of others, interpret their actions and reactions, and adjust our own behavior accordingly, thus facilitating smooth navigation through the complex web of human relationships.

The study of ToM is a vibrant and multidisciplinary field of research, attracting interest from various areas of knowledge. Developmental psychology investigates how and when this capacity emerges in childhood and how it is refined throughout life, examining the milestones and factors that influence its acquisition. Neuropsychology explores the neural bases of ToM, identifying the brain regions involved and the dysfunctions that can arise from injuries or neurological disorders, as in the case of individuals with autism. For its part, social neuroscience uses advanced neuroimaging tools to understand how the brain processes social information and how neural networks associated with the attribution of mental states are activated in interactive contexts.

In addition to these disciplines, ToM is also relevant to the philosophy of mind, which debates the nature of consciousness and intentionality, and to artificial intelligence, where the goal is to replicate this capacity in computational systems to develop more sophisticated and socially competent agents. Understanding ToM not only sheds light on the functioning of the human mind but also offers valuable perspectives for addressing challenges in the educational, clinical, and technological spheres.

The following comparative table summarizes the different levels of inference, organized by their complexity.

Level	Support	Type of Inference	Example	Explicit
0	Physical-Quantum	Energetic Resonance	Particles seeking states of minimum energy	No
1	Molecular-Cellular	Biochemical Communication	Quorum sensing in bacteria, immune evasion by probiotics	No
2	Ecological-Vegetal	Adaptive Signaling	Wild tobacco, corn emitting vola tiles	No
3	Metacellular (no brain)	Functional Manipulation	*Ophiocordyceps*, viruses modifying hosts	No
4	Implicit Cognitive	Basic Prosociality, Behavioral Adjustment	Rats, macaws, squirrels	Partial
5	Complex Social Cognition	Intentional Inference	Corvids, great apes	Yes
6	Symbolic-Linguistic Cognition	Metareflexive Representation	Humans	Yes
7	Simulated Artificial Cognition	Functional Modeling of the Other	Conversational AIs with simulated intentionality	Partial

Let's take the example of bacteria. Despite lacking a nervous system, they exhibit collective action facilitated by quorum sensing. This molecular communication mechanism allows them to sense the concentration of chemical signals in their environment. Upon reaching a population threshold, the bacteria coordinate the activation of specific genes for actions such as producing biofilms, releasing enzymes, or modifying their surroundings. This suggests that while there is no "mind" involved, something more than a simple reaction exists: an assessment of the population context and a coordinated decision.

Probiotics, masters of adaptation and intercellular communication, deploy a fascinating strategy to interact with their environment and, in particular, with the host's complex immune system. Their ability to modify the composition and structure of their cell envelope, a kind of molecular camouflage, allows them to evade detection and aggressive immune responses. This programmed "mutation" of their cell surface is not random; it is a strategic adaptation that allows them to present themselves in a specific way, thereby influencing the perception and response of the immune system.

In addition to modifying their envelope, probiotics are skilled emitters of molecular signals, a sort of biochemical language they use to communicate with host cells. These signals, which can include metabolites, enzymes, or cell wall components, act as messengers that modulate the activity of immune cells, altering their behavior and response patterns. It is as if the probiotics are "speaking" to the immune system, persuading it to adopt a more tolerant or less inflammatory stance. This sophisticated interaction, which combines altering their "presentation" with emitting specific signals, allows probiotics not only to go unnoticed but also to actively influence the immune system's perception of them and other microorganisms in the gut. By manipulating these "recognition signals" and "intercellular languages," probiotics manage to shape the host's immune response, promoting a balance that favors intestinal and overall health. In essence, by modulating their cell envelope and emitting biochemical signals, probiotics are functionally exercising a form of "modeling" of the "other"—that is, the host's immune system.

Let's move on to plants. In wild tobacco (*Nicotiana attenuata*), for example, it has been observed that the plant changes its flowering pattern depending on the type of pollinator and the risk of predation by larvae. Furthermore, it emits volatile compounds that attract natural predators of these larvae. It not only detects its aggressor but also summons a third party to neutralize it. Corn, on the other hand, when attacked, releases chemical signals that attract natural enemies of the attacker and are also detected by neighboring plants, which activate their defenses in anticipation. These systems of inter-plant and inter-species signaling do not imply a mind, but they do show complex adaptive capacity and ecological sensitivity.

In the world of fungi and viruses, we observe even greater sophistication. The fungus

Ophiocordyceps unilateralis infects insects and modifies their behavior, leading them to optimal positions for releasing spores. Viruses, in turn, hijack the host's genetic machinery and manipulate its cellular and immunological processes to maximize their replication. Some even promote cellular responses that allow them to remain latent. In these cases, evolutionary adaptation has given rise to true functional models of the other: of their physiology, their behavior, their defense system [Hughes2011].

In non-human animals, we find increasingly solid evidence of forms of social inference. Gray rats show prosociality: if they receive help—for instance, being freed from a trap—they will later help other rats, even if they are not the same ones [Rutte2007]. This type of indirect reciprocity implies social memory and a willingness to act for the group's benefit. Macaws, in controlled experiments, give tokens to companions so they can exchange them for food, even if they themselves get nothing in return [Peron2011]. This behavior does not seem to be trained or conditioned and suggests a representation of the other's situation.

13

Gray squirrels, for their part, re-hide their food reserves if they have been observed by other individuals, which suggests sensitivity to the attention of others and risk assessment.

The evidence for the existence of complex mental representations, including epistemic ones (knowledge about the knowledge of others), is particularly robust in great apes and corvids. These findings challenge the notion that theory of mind is an exclusively human capacity.

In chimpanzees and bonobos, the sophistication of their social cognitive abilities is astonishing. Not only are they able to pass classic false-belief tests, which implies understanding that another individual can hold an erroneous belief about reality, but they also demonstrate a much broader range of behaviors that require a deep understanding of others' mental states. For example, these primates have been observed to actively resort to deception and social manipulation. They can hide information, divert the attention of others, or feign intentions to achieve their goals. Moreover, they adjust their behavior flexibly and strategically based on what they perceive the other individual knows or is ignorant of. This is manifested in situations like competition for resources or the formation of alliances, where their success depends on their ability to predict and manipulate the actions of their peers based on their knowledge and beliefs.

Corvids, particularly crows, have demonstrated cognitive abilities that rival those of primates in certain respects. Their impressive spatial memory and ability to plan for the long term are well known. However, what truly underscores their theory of mind is their behavior surrounding food caching. Studies have revealed that crows not only remember where they hid their food but also who observed them during the caching process. If they perceive a risk of theft from those observers, or even from other crows that might have been watching, they relocate their food to safer places. This behavior cannot be explained simply as a reaction to the presence of a predator but suggests an understanding that the other individual possesses specific "knowledge" about the food's location and, therefore, the "intention" to steal it.

Together, these examples in great apes and corvids demonstrate the existence of mental representations that are not only functional—that is, they allow them to interact effectively with their social environment—but also epistemic. This means these animals possess the ability to attribute mental states like beliefs, knowledge, and intentions to other individuals and to use this information to guide their own behavior. These findings are crucial for understanding the evolution of social cognition and the complexity of non-human minds.

Humans develop this capacity around 4 or 5 years of age [Wimmer1983]. It is the basis of social interaction, language, empathy, and morality. It manifests in the ability to imagine what another person thinks, feels, or believes, even if they are wrong. It has been studied by developmental psychology (with the classic false-belief test), by neuroscience (with activation studies of the medial prefrontal cortex and the temporoparietal junction), and by anthropology (as the basis of cooperation and culture).

Artificial intelligence has begun to emulate some aspects of this capacity. Models like ToMnet, proposed by DeepMind, simulate inference about agents with different beliefs [Rabinowitz2018]. Conversational agents like current language models adjust their responses based on context, the user's apparent intention, and the conversation history. Although they have no mind or consciousness, they exhibit a functional theory of mind: they model the user as an entity with goals, partial knowledge, and communicative needs [Taleb2012].

But here a key question arises: Is that truly ToM, or a high-level simulation?. Does intent matter, or only function? Must there be consciousness, or is it enough to represent the other effectively?.

A non-interventionist documentary captured a moving scene: a young deer was struggling desperately, trapped in a small stream.

Despite its efforts, it could not get out. Its herd passed nearby, indifferent. Simultaneously, the cameras focused on an adult elephant observing the entire situation from a medium distance. The elephant did not react immediately. Its gaze lingered on the deer's herd. It emitted a short, sustained sound, waiting. Then— following the indifference of the passing herd—it turned its head toward the trapped deer in the stream, approached, and repeated the sound. With extreme care not to slip in the mud, it assessed the terrain. Extending its trunk with a controlled gesture, it grasped the deer by the torso, extracted it, and set it safely aside. The deer shook itself, stood up, and ran towards its group. The elephant, in silence, let out a short breath, shook its ears, and continued on its way.

This scene goes beyond mere instinct, imitation, or instrumental learning. It was not a member of its group, nor its offspring, nor a competitor. There was no obvious reward for its effort, nor any personal danger motivating it. Here, the key is something deeper: There is representation. There is inference. There is decision. And, perhaps—to anthropomorphize—there is compassion. The elephant not only understood the trapped deer's situation but also that of those who watched without acting. It understood they would do nothing. It was then that it acted, assuming the role of the only being conscious of all the minds present, including its own responsibility as the one who "could" help. This is not about trying to prove whether the elephant "believes that the deer believes it is trapped"; that framework is insufficient. The language of traditional cognitive science pales before the silent elegance of this gesture. But right at that moment, when a non-human animal seems to understand more than the others, is when we must introduce the concept we are missing: Theory of Mind.

Theory of Mind (ToM) has traditionally been conceptualized as a cognitive milestone, a discrete threshold that humans cross to understand and attribute mental states to others. However, a more nuanced perspective suggests that ToM is not an "all-or-nothing" binary phenomenon, but rather a continuous spectrum of abilities. This progression extends from the most fundamental, energetic interactions between subatomic particles to the intricate narrative inferences that characterize human social cognition. Theory of Mind should not be thought of as a single capacity that appears suddenly, but as an emergent cognitive function, distributed across different levels of complexity depending on the species, environment, and mechanism of inference.

Below are six possible levels, organized from functional behaviors without explicit inference to the complex simulation of others' mental states.

Level	Type of Inference or Sensitivity	Biological or Cognitive Support	Paradigmatic Example	Classic ToM?
0	Instinctive reaction to the other	Genetics, evolutionary reflexes	A newborn deer fleeing from danger	no
1	Contextual behavioral rules	Associative learning, plasticity	A squirrel re-hiding food after being observed	no
2	Ecological or chemical inference of the social environment	Chemical signals, quorum sensing, biochemical communication	Plants altering defenses based on neighbors or enemies, bacteria activating genes upon reaching critical density	"no, but functional"
3	Functional modeling of the other without symbolism	Implicit social cognition, perception of agency	Rats helping others after being helped, parrots transferring tokens to partners without personal benefit	partial
4	Explicit intentional representation within the species	Advanced cognition, simulation, strategic deception	Crows pretending to hide food, chimpanzees understanding false beliefs	yes
5	Interspecies representation of agency and suffering	Moral cognition, situated compassion, complex inference	An elephant rescuing a trapped deer while others ignore it	surpasses standard tests
6	Symbolic, narrative, and recursive mental simulation	Language, culture, introspection, metacognition	Humans writing fiction, creating characters, attributing layered mental states	Total ToM

Considerations: a) The levels are not mutually exclusive; the same species can exhibit behaviors at more than one level depending on the context . b) Classic ToM is located in levels 4–6, but the functionality of the lower levels is key to the evolution of the social mind . c) The intermediate levels (2 and 3) are the most overlooked in academic literature but are fundamental to understanding the transition between adaptive biology and social cognition.

In this continuum, the ability to model others, to predict or interpret their behavior, manifests in various ways. Not all levels of this ability require the presence of a "mind" in the traditional sense, or even a complex nervous system. Consider, for example, unicellular organisms that adjust their movement in response to chemical gradients in their environment, or insect colonies that coordinate their actions through pheromonal communication. These adaptations, while lacking the complexity of human cognition, exhibit a rudimentary form of "orientation toward the other," modifying their own behavior based on alterity.

The roots of these adaptations are diverse. Some emerge directly from evolutionary pressure, where the ability to anticipate or respond to the actions of others confers a survival advantage. Others develop through learning, both at the individual and cultural levels, refining the understanding of social dynamics. Even in the realm of artificial simulation, we see systems that develop the ability to model and adapt to the behavior of other agents, replicating certain aspects of what we might consider a computational ToM.

Despite their varied manifestations and origins, all these instances share a fundamental characteristic: they are intrinsically oriented toward the other. They involve an adjustment of one's own behavior in response to the presence, actions, or even implicit states of the "other," whether it be a particle, an organism, a conspecific, or an artificial system. This orientation to the other is the cornerstone of any form of social and adaptive interaction.

This expanded conception of ToM forces us to revisit a crucial and deeply philosophical question: Are we, in observing these manifestations in biological and artificial systems, simply projecting our own anthropocentric conception of "mind" onto them?. Or are we, on the contrary, unveiling a deeper truth: that the mind, as we know it in humans, is simply an extraordinarily sophisticated and complex manifestation of a fundamental adaptive resonance, a capacity inherent in life to adjust and respond to the environment and to other agents within it? This perspective suggests that what we call "mind" could be the culmination of a long evolutionary history of other-oriented interactions, a reverberation of life responding to life itself.

ATTENTIONAL SPOTLIGHT

The Key to Learning

Years ago, during my first days at university, my friend "El Checo"—the Czech—had already been married for several years. Driven by my usual curiosity, I asked him how he managed to stay faithful to his wife. He spoke of love, commitment, and all the predictable clichés, but in the end, he said something that stuck with me:

"I keep my eyes on what matters. When the sting of temptation threatens, I think about everything I would lose: my family, my home. It's not worth it. "

I wanted to provoke him a little more: "But isn't it okay to at least look at the menu?"

Without losing his composure, he replied: "Why would you look at the menu in a restaurant if you're not going to eat? When you're married, those things aren't just yours anymore; they belong to the couple."

Beyond the marital advice, what I was left thinking about was his ability to direct his attention. Because that's what this text is about: how the control of attentional focus—the ability to decide what to concentrate on and what to ignore—is far more powerful than we usually imagine.

Attention is the gateway to our consciousness. Everything we perceive, learn, and remember first passes through that filter. It is the lens through which we construct reality. From a neuroscience perspective, we know that attentional control is linked to specific brain regions, especially the dorsolateral prefrontal cortex [Miller2001]. This region, which continues to mature well into adulthood, is involved in decision-making, self-control, and emotional regulation. When we exercise voluntary attention or inhibit an impulse, this area activates and coordinates other brain regions, like the amygdala or the striatum.

In the late 1960s, psychologist Walter Mischel conducted a famous experiment at Stanford University. A group of children were offered a treat—a marshmallow, a cookie, something tempting—with two options: they could eat it immediately or wait about 15 to 20 minutes and receive a second one as a reward. The interesting part came later: Mischel and his team followed the lives of these children for more than a decade. Those who had resisted temptation tended to show higher self-esteem, better frustration management, and better academic performance as adolescents [Mischel1989].

But most revealing was what Mischel discovered by observing how the children who waited managed to resist. They didn't do it through sheer willpower. They used distraction strategies: some covered their eyes, others sang, danced, looked at the ceiling, or imagined the marshmallow was a cloud. Mischel called this "strategic allocation of attention," and he concluded that this ability—to voluntarily redirect attention—was the real foundation of self-control [Mischel1970].

Attention is not a single entity. A distinction is made between voluntary (endogenous) attention, which is intentional and directed from within, and automatic (exogenous) attention, which responds to environmental events like loud noises or striking images. Strengthening attentional focus involves training the ability to sustain endogenous attention against the constant seduction of external stimuli.

We also know that this skill can be trained. Neuroplasticity studies show that the constant practice of mindfulness exercises, for example, increases gray matter density in areas of the prefrontal cortex associated with emotional regulation and concentration

[Holzel2011]. In other words, directing your focus is not just a mental decision: it is also a physical change in the brain.

We often believe that willpower is a moral virtue, almost a matter of character. But what really makes the difference is the ability to direct attention: knowing how to manage that short list of thoughts floating in our working memory. Although it was classically held that this memory could only retain between 5 and 9 items at a time (Miller's "magical number 7") [Miller1956], recent research suggests this figure is more limited and flexible, closer to 4 items, and depends on the type of information, an individual's prior experience, and the context of the task [Cowan2010].

Decades later, science revisited Mischel's experiment. Some replication studies showed that the effect of childhood self-control on future success was real, though more modest than originally thought, especially when factors like family environment and socioeconomic status are considered [Watts2018]. However, this does not weaken the importance of attention. On the contrary, it reminds us that attention is a powerful tool, but not the only one. It is part of a more complex ecosystem of skills and circumstances.

The fascinating thing about this capacity is that it transcends childhood. Years after the original experiment, Mischel re-evaluated the same children, now adolescents. Those who had not resisted temptation at age four now had more difficulty maintaining friendships, concentrating in class, and regulating their behavior. Once again, attention and delayed gratification appeared as predictors of success and well-being [Mischel1989].

From finishing a demanding project to cultivating gratitude or resilience, the control of attentional focus is a master skill. It helps us prioritize, learn better, and think clearly in the midst of chaos. In this era saturated with stimuli, developing this skill becomes more crucial than ever. As the economist and Nobel laureate Herbert Simon said: "A wealth of information creates a poverty of attention" [Simon1971].

23

We live surrounded by notifications, social media, and algorithms that compete for every second of our gaze. This overload not only erodes deep concentration but also fuels anxiety, superficial thinking, and decision fatigue. Remarkably, these same principles find application in artificial intelligence. In computational models like transformers, the "attention" mechanism allows the system to selectively focus on the most relevant parts of a sequence of data, ignoring the rest [Vaswani2017]. Analogous to the human brain, this process does not treat all information equally but assigns "weight" to what matters most for the task. This suggests that directing attention is not only biologically adaptive but also computationally efficient.

Modern neuroscience has been dismantling the idea of absolute free will, revealing how many of our decisions are automatic, shaped by habits, biases, and unconscious mechanisms. However, in this complex web, the ability to direct our focus of attention emerges as one of the few and most powerful levers of conscious control we truly possess. It is in this small space of intentional control that our real power to change, learn, and evolve resides. By mastering this skill, we not only resist the "marshmallow" of distractions and immediate temptations but also become the true architects of our mind, laying the foundation for deeper learning and a more intentional and meaningful interaction with the complexity of the world around us, transcending mere reaction to actively build our reality and our future.

DOPAMINE AND MOTIVATION

Why do certain activities like video games or social media effortlessly absorb us, while others with greater long-term benefits—like studying or starting a business—feel like a constant struggle? Is there a way to make these challenging and sustainable tasks more appealing? The key lies with a tiny but powerful actor in our brain: dopamine.

Dopamine is a crucial neurotransmitter that influences motivation, learning, pleasure, and goal-directed behavior. It is fundamental to the brain's reward system, which is a network of interconnected brain structures responsible for processing pleasure, motivation, and learning. Its impact extends beyond basic biology, profoundly shaping our interactions with the world. To comprehend its function, we must understand its role within the brain's intricate reward network. In this context, dopamine acts as a chemical messenger, driving us to seek and repeat behaviors essential for survival, such as eating and drinking. Research, particularly in animal models, demonstrates dopamine's collaboration with other neurotransmitters. This cooperation modulates motivation, associative learning (the connection between an action and its reward), and emotional state—all processes vital for decision-making and the shaping of behavior [Salamone2002].

When we experience something gratifying, dopaminergic neurons release a burst of dopamine. This not only generates pleasure but also reinforces the neural connection between the action and the reward, encouraging us to repeat that behavior [Schultz1998].

Dopamine's powerful role in reinforcing behaviors makes it central to discussions about addiction. Although researchers agree on its critical involvement in the development and persistence of addiction, its precise function in humans is still being explored [Volkow2014]. Beyond initial euphoria, dopamine influences 'incentive salience' (which is a stimulus's capacity to grab attention and drive seeking behavior), even when pleasure decreases [Berridge1998]. Long-term exposure to addictive substances or activities can disrupt the dopaminergic system, leading to a compulsive pursuit and diminished ability to find pleasure from other sources [Koob2016].

The same power of dopamine that underlies addiction is clearly manifested in our relationship with digital activities, such as video games and social media. Neuroscience posits that repetitive activities activate the brain's reward system, generating spikes in dopamine. This effect, however, is not exclusive to the digital realm; any gratifying experience—from a delicious meal to a brief daily exercise or a compliment—can trigger this release, which is the brain's natural response . The amount of dopamine released varies significantly depending on the novelty, intensity, and unpredictability of the stimulus.

But let's turn our attention, for a moment, to video games. There is a cultural perception, based more on anecdotes than scientific evidence, that video games foster excessive leisure and laziness, diverting young people from "productive" activities. Particularly, subgenres like First-Person Shooters (FPS) and Action Shooters are often linked to aggression and social alienation. However, this relationship has been widely questioned. Various studies and meta-analyses have debunked this connection, showing no conclusive or significant correlation between time spent playing violent video games and an increase in aggressive behaviors [Ferguson2015]. In fact, most current studies suggest that video games are not inherently harmful when consumed in a balanced way [Granic2014].

The versatility of dopamine and its role in learning transcends mere entertainment, finding fundamental strategic application in education. Video games, for example, not only entertain; they are used constructively in behavioral therapies, retraining the reward system, and enhancing mental, physical, social, and creative skills [Granic2014]. They foster problem-solving, rapid decision-making, and collaboration. Current research seeks to clarify whether their reward systems motivate participation or are a result of the participation itself.

In the educational sphere, dopamine plays a crucial role in student motivation. Schools and training contexts worldwide are adopting pedagogical models that seek to activate the brain's reward system through engaging learning experiences. Approaches like Project-Based Learning (PBL), competency-based learning, the "Flipped Classroom," and experiential learning intrinsically aim to generate a cycle of reward and satisfaction.

Particularly relevant is gamification in education, where game-like elements—such as rewards, points, badges, levels, and competition—are integrated into the learning environment. This not only aims to make studying more fun and motivating but also increases engagement and interaction among students [Kapp2012]. By structuring the dosage of activities with clear academic objectives and by reinforcing achievements and positive behaviors (rather than focusing solely on negative consequences) dopamine release is optimized, and intrinsic motivation for learning is fostered.

Analogous to education, dopamine's impact extends to the workplace, where companies employ strategies to foster motivation and engagement in their teams. Agile methodologies like Kanban and Scrum, the establishment of clear goals, and business gamification are tools designed to activate the reward system. Whether through computerized systems or on a simple whiteboard, these tools allow for visualizing tasks, measuring progress, and providing immediate recognition for achievements. This positive reinforcement, which generates micro-doses of dopamine, maintains interest and dedication. The key to sustaining motivation lies in setting short-term objectives, or breaking down long-term goals into smaller milestones, celebrating each advance, and enjoying the continuous progress.

Beyond the specific environments of work and study, the principles of dopamine are intentionally applied in the design of products that seek to capture and maintain our attention and participation. Social media, video games, and even everyday products like a coffee service share this common goal. To achieve this, manufacturers and designers employ a blend of interdisciplinary mechanisms. In product design, Design Thinking, Design Research, and gamification stand out.

Design Thinking is a human-centered approach that seeks to understand the deep needs of users. Design Research, for its part, focuses on obtaining real data about their behaviors and expectations. Together, these approaches allow for the creation of more relevant and attractive products. Once launched, continuous user feedback allows for iteration and improvement of the experience. Even when a product is not a game, gamification incorporates elements from video games, such as incentives, achievements, and instant feedback, creating a cycle of expectation and reward [Werbach2012].

Video games, for example, use progressive difficulty with often unpredictable rewards to keep the player immersed. Similarly, social media exploits human social behavior through random notifications, thereby maintaining the attention of its users. These strategies seek not only to maintain attention but also to exploit neuropsychological mechanisms that induce the repetition of behaviors. However, it is crucial that this design is ethical and respectful of the user's well-being [Alter2017]. By integrating Design Thinking, Design Research, and gamification, organizations design highly motivating experiences tailored to the user's needs and desires, creating cycles of continuous improvement that are fed by both human behavior and constant feedback.

Despite concerns about negative effects like addiction, recent studies indicate that only a small fraction of *gamers* exhibit problematic behaviors [Przybylski2017]. In general, video games can contribute to the development of cognitive and social skills, enhancing collaboration and creativity.

However, the reward system is not static; its effectiveness can be altered by a phenomenon known as dopamine tolerance. When the brain is exposed to frequent and intense stimuli that generate rewards, the dopamine system can adapt. This process is not exclusive to exposure to digital environments or substances; it occurs with any activity that generates gratification. The adaptation can vary significantly among individuals. Indeed, numerous studies have shown that constant exposure to immediate gratification can cause certain activities to lose their appeal over time [Volkow2014]. This can lead many people to prefer activities that provide instant reward, making it difficult to focus on long-term goals that require sustained effort before a significant benefit is seen, as evidenced by studies on delayed gratification [Mischel1989].

Fortunately, this adaptation or "tolerance" is not irreversible; there are behavioral modifications that can help. The key is to implement "structure" and make small adjustments to your routine. This includes setting time limits for certain activities or creating routines that motivate gradual progress. For example, by transforming a large goal into smaller, achievable ones, more frequent dopaminergic rewards are generated, reinforcing the desired behavior. Reflect on how activities affect you—positively or negatively—and seek support if you need it. Small changes can lead to significant improvements in your self-regulation capacity.

But what happens when this vital system doesn't function correctly? Research also explores when dopamine fails. Studies indicate that the human body always produces some level of dopamine. However, certain conditions can cause a significant decrease in functional dopamine. These include Parkinson's disease, certain nervous system disorders, psychiatric disorders like severe depression or schizophrenia, and prolonged substance abuse (drugs or alcohol) that depletes or damages dopamine receptors [Dauer2003]. Understanding dopamine is, therefore, not

29

just about high performance—it's also about mental and physical health.

Dopamine is the invisible engine that drives our motivation, our learning, and, ultimately, our decisions, directly influencing what and how we fix our focus of attention. As we have explored from a neuroscience perspective, it plays an essential role in prioritizing our activities, actively guiding our mind toward what promises reward or meaning. Its power, however, comes with a responsibility: continuous and excessive exposure to highly gratifying stimuli, designed to exploit this system, can lead to an alteration of our reward circuit, driving us into a compulsive search for pleasurable sensations that, paradoxically, makes us less sensitive to them and hinders our concentration on meaningful long-term tasks [Koob2016].

To achieve sustainable personal and professional growth, it is imperative to learn to consciously manage our reward system. This involves breaking down large, long-term goals into smaller, manageable ones. Each time we reach one of these micro-objectives, our brain releases dopamine, reinforcing the path. The importance we give to each success, no matter how small, reconfigures how our brain anticipates and handles reward, improving not only our productivity but also our intrinsic ability to find satisfaction in progress.

Understanding dopamine equips us with a powerful tool to design environments and habits that, instead of draining us, empower us to build the life we desire. And in doing so, we are applying a fundamental principle of our own brain: the construction of the new always rests on the foundations of what we already know, weaving new connections into the intricate web of our prior knowledge.

PRIMED TO LEARN

How Evolution and Experience Prepare the Mind for Learning

Restless, a young kingfisher balances on a branch extending over the river. With its gaze fixed on the moving water, it searches for anything: a passing shadow, a flash of light, or a dry leaf floating by. Suddenly, with a precise and instinctive movement, it dives into the water. Its plunge is flawless. However, it catches a leaf instead of prey and returns to the branch, where it shakes it vigorously, again and again, before letting it go and starting over. It is not playing, but rehearsing. It practices with what the river offers, preparing for the real hunt.

This gesture, which seems like a game, holds a fundamental truth about learning: the new can only occur if something prior was already in place to receive it. That kingfisher's rehearsal is possible because there is prior knowledge driving it. What follows, in all species, is a dance between that inherited knowledge and an openness to change. This mechanism of trial and error, of linking a new action to a pre-existing schema, is the basis of all cognition. Theorists like Jean Piaget would describe it as the interaction between assimilation (incorporating new information into existing schemas) and accommodation (modifying those schemas to adapt to the new information) [Piaget1952].

The kingfisher does not need to be taught how to hunt, just as a spider does not attend classes to weave its first web or a newborn deer stands up within minutes of birth to escape danger. This is primordial knowledge, a phylogenetic wisdom encoded in DNA, inscribed in the architecture and chemistry of cells [Lorenz1965]. It is a map that each organism receives at birth, a memory of the species that precedes all individual experience. This knowledge is a functional and automatic response, processed in the most primitive regions of the brain, such as the brainstem and the limbic system. It is accumulated information available in our cellular networks—systems perfected over the evolution of species, designed to ensure survival and provide a starting point before experience has time to teach anything.

Similarly, a young arctic fox does not need an instructor to learn how to hunt under the snow. It is born with the instinct to listen for the movement of rodents beneath the white layers. However, the fox kits spend hours, even days, practicing their 'plunges': jumping and driving their heads into the snow. This apparent initial clumsiness is, in fact, a vital rehearsal. Although the hunting instinct is innate, the precision of the 'plunge' is perfected through repetition and feedback from the environment, transforming the genetic predisposition into a polished skill. Each attempt, whether failed or successful, refines the neural circuits of that inherited action, preparing the fox for effective hunting.

But the map is not the territory. True learning is something else; it demands openness and a desire that, in psychological terms, we call intrinsic motivation [Ryan2000]. When an organism feels curiosity, need, or interest—when a task is inherently rewarding—its biology changes. The brain releases key neurotransmitters like dopamine—that fundamental engine of the reward and motivation circuit we have already explored in detail, which drives us to repeat pleasant or meaningful actions—and acetylcholine, essential for focusing attention and consolidating new memories. These molecules act as key modulators of synaptic plasticity, preparing the neural circuits to reorganize and signaling that something in the environment is important enough to deserve an update of the map [Hasselmo2006].

Imagine our brain is like a closet with countless drawers. Every time we learn something new, we store that information in a specific drawer. If we want to learn more about that topic, it is much easier if we already have a "drawer" dedicated to it. That drawer, filled with prior knowledge, provides us with a framework to understand and connect the new information. That is, learning is a continuous mental process that manifests in our actions and behaviors, generating lasting changes in our knowledge, beliefs, and conduct. It is formed from how we interpret and respond to our experiences, both conscious and unconscious, always anchored in that pre-existing infrastructure. For example, if we already have a basic knowledge of European history, it will be much easier for us to understand an article about the French Revolution; our prior knowledge activates the relevant neural circuits, allowing us to make connections and understand the context of the event with greater fluency.

From our human perspective, we call this process attention, interest, or the desire to learn. It is the biochemical spark that ignites neural plasticity and opens the door to change. More than a lesson from our ancestors, it is the activation of inherited motor predispositions and an implicit procedural memory—a know-how that does not require consciousness but serves as scaffolding for the new skill. Thus, prior knowledge is not just a repository of facts, but a set of already active and reinforced neural pathways that facilitate the assimilation of new information, making the learning process more efficient and less cognitively costly.

However, prior knowledge is not always an ally. Sometimes, it can hinder our learning. This occurs when the information we already have is incorrect, incomplete, or irrelevant to the current situation, or when it is structured rigidly. A common example is stereotypes: if we have preconceived and erroneous ideas about a group of people, we may misinterpret their actions and behaviors, which hinders our genuine learning about them. These pre-established neural pathways, instead of facilitating, act as barriers, filtering new information in a biased way.

In addition to accuracy, the way we organize that information is crucial. If our "mental drawers" are disorganized, we will struggle to find what we need, and new connections will be difficult to establish. On the other hand, if we have a clear and structured system—a system of well-interconnected and accessible neural networks—we can access our prior knowledge efficiently and use it as a solid foundation to build new learning. The organization of knowledge is, in essence, the optimization of our neural highways for the flow of new information.

But this openness to new knowledge can be drastically altered if the brain interprets certain environmental signals as threats. A classroom, for example, could be perceived as full of 'hostilities', from the temperature or lighting to a severe look, a humiliating correction, or the expectation of failure. In such circumstances, the brain activates an ancestral 'fight or flight' response, designed for immediate survival and not for learning [LeDoux1996]. The amygdala, the brain's threat detector, orchestrates the activation of these defense circuits, releasing hormones like cortisol. This prepares the organism to fight, flee, or freeze, redirecting energy and cognitive resources from learning and memory functions toward physical readiness. Theorists like Knud Illeris describe this phenomenon as a 'defense against learning', where the organism actively protects itself from new information perceived as dangerous [Illeris2007].

Neuroscientifically, this state of high alert or 'amygdala hijack' inhibits the activity of the prefrontal cortex, the area responsible for executive thinking, planning, and complex learning. In this state, the cellular networks that could reorganize to learn—thanks to the chemistry of motivation and the production of electrical impulses between neurons—shut down like a flower at night. No dopamine is generated, no connections are reinforced; learning stagnates. Even prior knowledge, besides the fact that it can work against us by activating old biases or fears, does not necessarily transform into learning if there is no biochemical openness to allow it.

This tension between what we already are and what we could become is at the very heart of learning. And intelligence lies not in the perfection of the initial map nor in the ability to explore the territory, but in the fluid dance between both. It is the ability to trust instincts when necessary and, at the same time, have the flexibility to rewrite our responses when the world changes. Understanding this is fundamental. Environments that foster learning—whether in

34

a classroom, in therapy, or in our personal lives—are those that promote psychological safety. They are spaces where the threat system calms down and the biology of curiosity can be activated. When our brain satisfies what Richard Ryan and Edward Deci's Self-Determination Theory identifies as the three universal basic psychological needs—the need for Autonomy (feeling our actions are voluntary), Competence (feeling effective), and Relatedness (feeling connected and cared for by others)—it interprets that it is safe to lower its guard and open up to the new [Ryan2000].

The kingfisher, rehearsing with its leaf, and the arctic fox kit, perfecting its plunge in the snow, show us in their most primary essence how, even in nature less polished by abstract cognition, knowledge is organized to make the most of it. Because to build a strong and effective bridge of knowledge, the ability to organize it is as vital as the knowledge itself. This organization is what determines our capacity to remember, apply, and adapt what we know to the challenges we face, thus optimizing the work of our "inner architect."

ORGANIZING KNOWLEDGE

The Architecture of an Effective Mind

Have you ever wondered why certain ideas stick in your mind with ease, while others, no matter how many times you repeat them, seem to vanish? The key lies not just in the amount of information we possess, but in the architecture with which we organize it. This essay delves into how prior knowledge, when effectively structured, becomes the foundation of our ability to remember, apply, innovate, and adapt.

Just as a kingfisher practices its dives with leaves before catching real fish, or the arctic fox rehearses its pounce on the snow to reach unseen prey, we too depend on organized patterns of knowledge to act effectively. Nature shows us that cognitive efficiency arises from an internal order. Even the *Mimosa Pudica*, which stops closing its leaves in response to a harmless stimulus after several repetitions, demonstrates how filtering, prioritizing, and organizing information is a universal strategy for saving energy and adapting better [Gagliano2014].

Imagine your mind is like a closet. Every piece of data you learn is an item of clothing. If you toss items randomly into the closet, finding the right one at the right moment will be chaotic. But if you organize the clothes by type, color, or season, you not only access them quickly but can also combine them creatively. The same happens with knowledge. When knowledge is disorganized, it is not only hard to retrieve but also difficult to make useful connections. The mind becomes a noisy and confusing space. But when information is ordered hierarchically, with explicit links between concepts, principles, and experiences, learning becomes fluid and powerful.

Research in cognitive science, such as that by Ambrose and colleagues, shows that experts not only know more, but they organize their knowledge into rich, connected conceptual structures that allow them to retrieve relevant information more easily [Ambrose2010]. They think in terms of systems, patterns, and causal relationships. Novices, in contrast, tend to accumulate isolated facts without clear connections. This phenomenon has been described as the difference between a well-classified library and a disorganized pile of books. The good news is that anyone can learn to build a more orderly system. In doing so, we not only improve memory but also creativity and problem-solving ability.

To achieve this, it is recommended to:

Connect: Establish relationships between concepts. How does this relate to what I already know? What part of my knowledge network is activated by this new information?

Structure: Group ideas into categories. Use outlines, mind maps, or hierarchical lists that show how concepts are grouped and ranked.

Visualize: Graphically represent what is being learned. Concept maps and diagrams help to solidify connections that are sometimes not evident in linear text.

Practice: Reinforce what has been learned through application, active recall, and spaced practice. Strategic repetition strengthens connections and reinforces the accessibility of knowledge.

These principles apply beyond the academic realm. Cooking, speaking a language, managing teams, or learning to play an instrument require the same ability to integrate and structure knowledge. And every time we manage to connect pieces of information fluidly, our brain releases small doses of dopamine, reinforcing the habit and making it pleasurable [Biederman2006].

Even when we are not actively studying, our brain organizes knowledge. The default mode network (DMN), active when we are not focused on specific tasks, is responsible for archiving and consolidating information, planning for the future, and reorganizing experience. Studies like those by Gusnard and Raichle have shown that this brain network is activated during states of mental rest and plays a key role in the consolidation of autobiographical memory and the simulation of the future [Gusnard2001], integrating past experiences with imaginative projections [Deco2011]. Sometimes, a solution arrives during a walk or a shower: it's not magic, it's the brain organizing its archive in the background.

This organizational capacity does not occur in a vacuum. Prior knowledge is the foundation. But its power depends on how it is stored. If a neural network contains well-connected and hierarchical nodes, it will be easier to integrate new information. If, on the other hand, it is a disorganized swarm, the new information will be lost among the old. That is why learning is not just acquiring: it is reconfiguring.

And this reconfiguration needs the right conditions. Stress, threat, or emotional pressure blocks the system's openness to learning. As we explained before, when the amygdala detects danger, it activates a series of biological responses—known as the FFF heuristics: 'Fight, Flight, Freeze'—that prioritize immediate survival [LeDoux1996]. This set of responses has evolutionary priority and temporarily displaces complex cognitive processes such as reflection, planning, and learning. Under these conditions, the brain interprets that learning is a luxury it cannot afford until the environment is safe. In contrast, environments where threat is minimized and curiosity or intrinsic motivation is activated facilitate the brain's opening of its networks to the reorganization of the mental archive. The organization of knowledge, then, is not just a matter of cognitive techniques, but also of affective conditions. In

38

this process, so-called mental models play a fundamental role: internal structures that we use to interpret information, anticipate outcomes, and make decisions [Johnson-Laird1983]. Their effectiveness depends largely on how they are organized, and they can either facilitate or hinder learning depending on their flexibility, coherence, and capacity for updating. Therefore, understanding and refining our mental models is an essential part of cognitive and emotional development.

In summary, prior knowledge is the raw material, and the organization of knowledge is the architecture that gives it shape and functionality. Together, they build the bridge to deep, flexible, and lasting learning, allowing us to not only store information but also transform it into applicable wisdom. By becoming conscious architects of our own minds, we learn to build and rebuild with meaning, purpose, and vision. It is in this domain of internal order that the true capacities to innovate, adapt, and find functional beauty in the complexity of life resides.

PLAY AS AN ADAPTIVE STRATEGY

The other day, tired from my morning jog, I stopped and happened to see a dog wagging its tail. It was playing—or so it seemed—with a cat, a pigeon, a crow, and a squirrel. The dog would playfully chase the pigeon, while the squirrel rode on its back and the crow ran in circles after the cat. And then they'd repeat the whole sequence. For a moment, I thought I was witnessing some kind of collective interspecies assault on the poor dog. But then I looked closer: the dog was wagging its tail, an unmistakable sign of its playful intent. This scene, bizarre at first glance, presents a daily miracle of wordless cooperation—a momentary utopia between species where, for a few minutes, the impulse to explore triumphed over the impulse to compete. It got me thinking: why do animals play? And if play is so fundamental in nature, why do we—as animals ourselves—tend to relegate it to childhood or dismiss it as a "waste of time" in adulthood? Is this unusual orchestra of species just a whim of nature, or does it hide one of life's most profound adaptive strategies: play?

The answer begins with everyday examples like the dog: what we observe as play is often a sophisticated adaptive strategy. Consider the young arctic fox, which, with its innate *'prior knowledge'* of how to hunt rodents under the snow, spends hours and days perfecting its 'plunges.' These jumps and head-first dives into the snow, though not always resulting in a real catch, represent a form of play that hones the precision of its instinctive behavior. Similarly, the young kingfisher rehearsing its dives with floating leaves is not wasting time; it is refining its hunting technique in a safe environment, organizing its hardwired knowledge to apply it effectively when real prey appears.

Furthermore, in the oceans, dolphins engage in a fascinating variety of games: from chasing air bubbles they create themselves, to "surfing" the waves made by boats, or even "jellyfish tossing," where they launch jellyfish to one another. These behaviors have no immediate reproductive or nutritional purpose, yet they develop agility, coordination, and social cohesion—crucial skills for cooperative hunting and the pod's social structure. In all these cases, both young and adult animals actively participate, showing that for animals, play is not an age-limited activity but a vital, lifelong laboratory for survival and behavioral refinement.

What we know is that play is widespread among animals, both domestic and wild, and appears in an incredible variety of species. Biologists like Marc Bekoff and Gordon Burghardt, who have extensively synthesized research in the field, have proposed criteria for identifying animal play: it lacks any apparent immediate benefits, is done for its own sake because it seems enjoyable, and uses "serious" behaviors (like fighting or hunting) but in an exaggerated, jumbled, or self-handicapping way [Bekoff2001, Burghardt2005]. Most intriguingly, play happens when the animal feels safe and free from threats. Flies chasing marbles, octopuses entertaining themselves with jets of water, dogs pretending to be clumsy. If play were useless, evolution would have eliminated it millions of years ago.

Emotional expression in animals varies greatly and is often misinterpreted. In dogs, a wagging tail indicates joy, while a rigid or rapid movement can signal excitement or aggression. Studies show that wagging to the right is associated with positive emotions, and to the left with negative ones, suggesting a brain asymmetry similar to humans' [Quaranta2007]. This movement also spreads pheromones. In cats, tail language is more subtle: an upright tail is a greeting, a rapidly twitching tail is irritation, and a puffed-up tail is fear. Cats are less social than dogs, and their tail movement is often a warning rather a sign of joy. The complexity of their communication, while different, is fascinating.

41

But let's get back to play. How can we understand the ubiquity of this seemingly useless activity? To start, we might have to rethink our most basic assumptions: maybe animals are faking it. And if they're faking it, then there's something like creativity involved—inventiveness. The dog wagging its tail isn't just happy; it might also be saying, "Relax, this is just a drill." That tail is a metacommunicative signal: a gesture that accompanies the action and adds a layer of meaning. It's like putting invisible quotation marks around the behavior—"this is not real, but let us play as if it were." These signals are present in many species: they indicate that the bite isn't real, the chase isn't a hunt, the roar is just for show. They are not mistakes or confusion, but rather shared constructs, a consensual fiction. In other words, play is a form of theater. And that changes everything. It forces us to think that animal behavior isn't always governed by the Darwinian logic of optimization—maximizing resources, minimizing energy, surviving at all costs. And then the question arises: Do cells or bacteria play? Are they the serious characters in the system, the austere accountants measuring every expense? While defining "play" in such simple organisms is a conceptual challenge, the fact is that all biological systems share fundamental characteristics built during evolution. The programming is shared: we all, in some way, run the same adaptive code.

Traditional theory suggested that animals play to practice behaviors necessary for adult life. However, recent studies indicate that play serves to experiment with and vary behaviors. Wild animals activate motor patterns non-repetitively, allowing them to test things out without risk. This functions as a laboratory where they spontaneously mix movements and strategies. In rats, it has been shown that social play in infancy develops brain areas associated with decision-making in uncertain situations [Pellis2010]. Play doesn't train a specific response, but rather the ability to navigate uncertainty and flexibility as an evolutionary advantage.

We rarely consider plants when thinking about play, often relegating them to the background due to their apparent stillness.

However, if play is understood as a form of exploration and adaptation, could plants also exhibit playful principles in their environmental interactions?

Thinkers like Stefano Mancuso, a pioneer in plant neurobiology, urge us to rethink our assumptions [Mancuso 2015]. He and other researchers have revealed plants to be incredibly complex organisms, far from passive, displaying sophisticated behaviors. While lacking a centralized brain like animals, their decentralized systems constantly detect and monitor numerous environmental factors. They engage in extensive communication with other plants, even distinguishing between relatives and non-relatives, and manipulate animals for their own benefit.

In "The Nation of Plants," Stefano Mancuso challenges animal-centric hierarchies, advocating for a vision of "broad and decentralized plant democracies" [Mancuso 2019]. This perspective compels us to view plants not as mere objects, but as "sovereign subjects."

To further illustrate this adaptive complexity, let's recall the fascinating case of wild tobacco (*Nicotiana attenuata*), which we've explored before. This plant has developed a sophisticated "dance" with the Sphinx moth and the hummingbird to ensure its pollination. When attacked by larvae, the plant not only modulates its floral behavior but also activates a chemical defense: the production of secondary metabolites. These compounds, like nicotine, not only make the leaves less palatable to caterpillars but also emit volatiles that attract the larvae's natural predators [Kessler2001]. This chain reaction transforms the initial threat into a complex "game" of signals and counter-signals that ultimately benefits the plant, demonstrating a surprising capacity for "negotiation" and "adaptive response" within the ecosystem.

And some flowers go even further in their strategies. Orchids of the genus *Ophrys*, for example, have developed an even more subtle deception. Instead of investing resources in nectar, they have perfected a disguise. They mimic the colors, shapes, and

43

even the scent of female bees. Their deception is so convincing that males, confused, copulate with the flower, and in the process, transport pollen to another flower of the same type. While the idea of plants "playing" in the animal sense is difficult to prove, research into their intelligence and communication opens the door to deeper reflection. Is it possible that the exploration of possibilities, the flexibility, and the adaptability that are at the heart of animal play manifest in plants in ways we do not yet understand?

Let's return to animals. In play, animals experiment with behaviors that in real situations could be ineffective or even dangerous, but without taking a real risk. The playful situation is a kind of "make-believe": a space where what happens is not literal, but a safe simulation. If real life were always perceived as unsafe or threatening, play simply wouldn't exist. What's fascinating is that during this "make-believe," animals seem to be trying new things, exploring possibilities outside of their usual functional routine. From a computer science perspective, this dynamic is directly linked to creativity theory, which identifies three fundamental processes: a) exploring an existing framework, b) creating new combinations of elements, and c) transforming the space of possibilities [Boden2004]. That's where play becomes a ludic experience: it is a dynamic laboratory for exploration and experimentation.

This phenomenon not only occurs in play within the same species but is particularly noteworthy in play between different species. In that case, the playground is also a space for communicative negotiation and shared creativity, where behavioral flexibility and invention become tools for coexistence.

But are we anthropomorphizing? How can we be sure that animals really play, that they invent, that they "pretend"? These are legitimate questions. The truth is, although we cannot "enter the mind" of another being, we can study its behavioral patterns. And with increasingly sophisticated tools, we have found consistent evidence that play is not a mere instinctive reflex, but an intentional, flexible, and socially communicative activity [Bekoff2001, Burghardt2005]. This leads us to ask: if animals play to explore possibilities, what does that say about our own need to play?

In fact, numerous studies have shown that video games not only entertain but also promote learning and improve cognitive skills like attention, memory, and problem-solving [Granic2014]. This powerful ability of digital play to facilitate the safe and structured exploration of new strategies has transcended the personal sphere, giving rise to business, educational, and social models based on gamification [Kapp2012]. Thus, play has established itself as a central tool not only for fun but also for innovation and training.

But here a sharp question arises: what about video games that simulate violence? Are players learning aggressive behaviors, or does the brain already have an implicit "contract" that allows it to differentiate fiction from reality? Science tells us that the brain is incredibly skilled at distinguishing fiction from reality. Just as we enjoy a horror movie without believing we are in real danger, our brain activates specific mechanisms, especially in the prefrontal cortex, that act as a tacit "contract" with the fiction [Mar2011]. In the case of violent video games, the research is complex. While exposure to violent content can have short-term effects, there is no conclusive evidence that playing violent video games makes people aggressive in real life [Ferguson2015]. In fact, most studies suggest that play, even in combat simulation contexts, can function as an outlet, a safe space to explore behaviors without real consequences.

Interspecies play and the practice of learning reveal, among other evolutionary heuristics, Brain Plasticity, Pattern Recognition/Learning, Theory of Mind, the "Use It or Lose It" principle, and Anticipation and Planning—all evolutionary heuristics that prepare the organism for future challenges at low risk by refining its movements and strategies.

In summary, from the biochemical defenses of a plant to the complex social simulations of dolphins or self-regulated learning in humans, play and error-adaptation strategies are not mere coincidences. They are manifestations of deep evolutionary principles: trial and error as the engine of evolution, the search for energy efficiency, plasticity, and the ability to transform adversity into a source of strength. Understanding this shared legacy allows us to appreciate the profound intelligence inherent in life itself.

Play, then, is not an obstacle to be eliminated, but a door to be crossed—a risk-free laboratory for training the mind and behavior. By embracing this playful heritage, we don't just survive: we thrive, building a future where flexibility and creativity are the true keys to life.

DO WE LEARN FROM OUR MISTAKES?

Why do we fail to learn from certain kinds of errors?

"You learn from your mistakes." This maxim is so deeply ingrained in our culture, repeated in motivational speeches and success stories, that we regard it as a universal truth of the human condition. We hear it from childhood, at school, and at work. And yet, if this folk wisdom is so true, why do we find it so hard to learn from our own errors? We need only look at our own lives—or at the great Thomas Edison, who famously said, "I have not failed. I've just found 10,000 ways that won't work"—to understand that learning from failure isn't the automatic process we often imagine. His ten thousand failed attempts, far from being obvious lessons, underscore the arduous nature of extracting valuable knowledge from each stumble.

Do we really learn from mistakes inherently? If we look at the course of our daily lives, this claim doesn't always hold up. Beyond the huge lessons that come from spectacular failures, why do we so often make the same mistakes over and over, ignoring warning signs that, in hindsight, are painfully obvious? What cognitive and emotional mechanisms—deeply rooted in our biology and psychology, and forged by evolution to ensure our survival—prevent us from capitalizing on every misstep, big or small, as a valuable lesson for the future?

The fundamental problem is that failure, especially when we perceive it as personal, can severely block or distort the learning process. When we make a significant mistake, a raw vulnerability emerges; the feeling of having failed, of not being good enough, can wash over us. To mitigate this discomfort, the brain deploys self-protection strategies. According to Leon Festinger's theory of cognitive dissonance, people experience discomfort when their beliefs (e.g., "I am competent") conflict with evidence of failure ("I made a foolish mistake") [Festinger1957]. To reduce that dissonance, the mind might rationalize, deny, or even actively distort information, rewriting the narrative of the error to protect the ego and prevent objective processing.

Beyond this, the brain can get stuck in what psychologist Susan Nolen-Hoeksema identified as **rumination**, a recurring thought pattern central to depressive and anxiety disorders [Nolen-Hoeksema2000]. In modern neuroscience, rumination is linked to the activation of the brain's default mode network, which is involved in self-reference and emotional evaluation. This creates a persistent thought loop about the same topic, failure, or negative consequences. This mental cycle, far from helping us find closure, keeps us anchored to the traumatic event or error, preventing new information from being constructively assimilated. It's a primal defense strategy, driven by mechanisms like ego-defense—an evolutionary trait designed to protect our identity and psychological stability—as well as the biological need to regulate our emotions.

In an attempt to preserve what's left of our self-esteem, the mind may resort to denial, rationalization, or projection. While the concept of projection was originally described by Freud, more recent studies on the **self-serving bias** support the idea that the brain tends to attribute failures to external factors to preserve self-image [Mezulis2004]. In this altered state, relevant information about the error's cause or potential solutions can be distorted, minimized, or ignored altogether, preventing any real progress. Paradoxically, while a car accident or a natural disaster immediately grabs our attention and compels us to find explanations—unpleasant situations often have a special salience that triggers our emotional memory and a bias toward the negative [Rozin2001]—when it comes to our own failures, we tend to get distracted and misperceive information, sabotaging our own chance to learn and grow.

There is a crucial threshold in our susceptibility to failure—a barrier that determines whether a mistake becomes a lesson or is simply ignored. If the outcome of a failure doesn't pose a real, imminent threat to our survival, well-being, or identity, or if the threat is perceived as "low-stakes," people are astonishingly inclined to ignore those missteps. Our brain, in its constant search for efficiency and "energy conservation," tends to simplify information and minimize cognitive effort.

For example, we don't perceive much danger in repeatedly forgetting our keys, not knowing the answers to a few questions on a "minor" exam, or leaving the car lights on again and again without immediate consequences. These everyday mistakes are too small, too insignificant, or too easy to rationalize as "slip-ups" for them to reach the level of emotional or cognitive impact needed to trigger deep learning. They become part of the background noise of our lives, easily blamed on bad luck or external factors rather than being analyzed as opportunities to adjust our own behavior. An even more common example is procrastination: we know that putting off an important task can have negative consequences, but if the deadline is far off or the immediate penalty isn't severe, we tend to ignore the 'error' signal in our planning, and fall into the same pattern repeatedly.

A telling example comes from medical residents who, during their first clinical rotations, make minor errors in technique with no serious consequences. Many of these mistakes don't trigger an immediate corrective response unless the outcome directly affects a patient or is flagged by a supervisor. In contrast, when the same error occurs with a real patient who experiences a visible—even if mild—adverse consequence, the emotional impact is usually enough to cement the lesson and change behavior.

On the other hand, if a mistake is of considerable magnitude and its consequences cannot be ignored without palpable harm—be it physical, emotional, professional, or financial—our attention is activated immediately and inescapably. This is the well-known psychological phenomenon of **"aversive learning"** [LeDoux1996].

Examples abound in neuroscience and behavioral research: lab rats vividly remember a particularly aversive event, like tasting poison or receiving a significant electric shock, and modify their future behavior to avoid that stimulus. Similarly, humans learn drastically and durably from their own experience when a failure has been forceful enough and the consequences unavoidable. In these high-impact scenarios, learning isn't just a correction but a robust restructuring: the memory of the event becomes almost indelible, and the change in behavior is swift and persistent.

In these high-stakes situations, **brain plasticity** is activated at an accelerated rate, as demonstrated by Eric Kandel's studies on long-term memory consolidation through synaptic changes [Kandel2001]. Exposure to intense, emotionally significant, or unexpected events stimulates structural and functional reorganization in key brain networks. This facilitates adaptive learning, allowing us to readjust our expectations and behaviors. Our resilience and adaptability are tested and even strengthened, as our very survival depends on recalibrating our actions.

The central question, then, is not whether we are inherently capable of learning from our mistakes, but under what specific conditions that learning is activated constructively and lastingly.

The brain is designed to find patterns and create mental closure to give our experiences coherence. But when a mistake is perceived as a direct threat to our self-worth, or when it's too insignificant to be processed as a relevant event, our automatic defense mechanisms or our cognitive-miser tendencies take over, nullifying the opportunity to learn.

Learning from everyday mistakes, therefore, is not a passive or automatic process that happens simply by having a negative experience. It requires an active **motivation to resolve what's incomplete**—a kind of curiosity or need for cognitive closure that drives us to solve the underlying problem. It also demands **metacognition**: the ability to reflect critically on the failure, to depersonalize it (that is, to see the error as objective data about a process, rather than a judgment on one's own worth), and to integrate that new information into our mental models [Fleming2012]. This transcends mere emotional reaction or avoidance.

Understanding these inherent barriers in our evolutionary and psychological wiring is the first critical step toward turning them into real opportunities for growth. Only by recognizing and disarming these built-in defenses can we transform every stumble into a stepping stone toward a more adaptive intelligence and a more fulfilling life.

SURVIVE, LEARN, AND THRIVE

Somewhere in a corner where nature deploys its most subtle strategies, a wild tobacco plant fights a silent battle for survival. Every night, its white flowers exhale a sweet perfume, an ethereal invitation to the tobacco hawkmoth. The moth, in its ancestral dance, feeds on the nectar and, unintentionally, carries the pollen that ensures the plant's genetic diversity. However, this pact carries a shadow: the moth lays its eggs, and from them emerge voracious larvae that threaten to annihilate their host.

When the caterpillars' appetite becomes unsustainable, the plant deploys an unexpected adaptation. Its nocturnal flowers close, reopening at dawn for a new visitor: the hummingbird. This new pollinator doesn't bring the larval threat, but neither does it offer the genetic richness of the wandering moth. It's a strategic sacrifice: a lesser benefit in exchange for immediate survival.

But the story doesn't end there. The presence of the hummingbird attracts other birds, who discover a food source in the caterpillars. Thus, the plant, through an adaptation to avoid a fatal error, triggers a cascade of interactions that benefit it in unforeseen ways. The threat becomes bait for a form of biological control.

This small ecological drama encapsulates the essence of what we can call the **Adaptive Error Thesis**: the idea that error, failure, and adversity are not mere evolutionary accidents or dysfunctions to be avoided at all costs, but necessary and often decisive mechanisms for the survival, reorganization, and innovation of complex systems. In this framework, an error ceases to be a flaw and becomes a signal, a lever that drives new adaptive paths. In nature, in the human mind, and in our organizations, failure is not simply what happens when we fall short; it is a critical source of structural information that we can metabolize to reconfigure our paths of action. Is this small ecological drama just a curiosity of nature, or does it encapsulate a fundamental truth about how life itself learns and thrives through error?

Error and Evolution

In the vast laboratory of biological evolution, innovation doesn't occur as a succession of correct moves but as a constant dance between what works and what is discarded as unfit. What we see today as successful life forms is the result of millions of years of failures, of non-adaptive mutations, of evolutionary paths that ended in dead ends. And yet, from this history riddled with what we might call failures, today's biodiversity emerged.

Natural selection, that great evolutionary crucible, does not reward perfection but rather contextual fitness [Darwin1859]. A mutation that seems beneficial in one environment may be meaningless in another; a defense strategy that saves a plant in one ecosystem could be lethal in another. Learning, in this sense, is not exclusive to organisms with brains: bacteria, plants, fungi, and even viruses respond to their environments with a trial-and-error capacity that allows them to adjust their behaviors, reconfigure their biochemical responses, and modify their life cycles. For example, a virus mutates to evade its host's immune system, "learning" to persist through its genetic mistakes. This implies that adaptive error is not an exclusively human phenomenon, but a universal principle that structures life.

53

Evolutionary Heuristics

This evolutionary dance didn't just shape our bodies; it hardwired our very minds with a set of mental shortcuts, or heuristics, designed for survival. In the human brain, these ancient predispositions manifest in several ways. We see it in the **fight-or-flight response** [Cannon1932], an automatic alarm system that, while useful in the wild, often misfires in response to a failed presentation. It's also present in our **negativity bias**, a tendency to fixate on criticism while glazing over praise [Rozin2001]. This is compounded by our innate **loss aversion**, which makes the fear of failure far more powerful than the hope of gain [Kahneman1979], and the paralyzing loop of **rumination**, which traps us in the past instead of letting us learn from it [Nolen-Hoeksema2000].

These heuristics are not design flaws in the brain; they are adaptations that worked in contexts where survival was the priority. However, in today's world—more abstract, symbolic, and fast-paced—they can become barriers to innovation, learning, and emotional self-regulation.

Strategies for Dealing with Error

Nature doesn't rely on a single strategy to adapt to error. Instead, it has developed a diverse portfolio of responses. At the simplest end of the spectrum lies **Fragility**, where a system collapses under stress. Though seemingly negative, this serves an evolutionary purpose by quickly weeding out non-viable solutions. A step above is **Robustness**: the ability to withstand disruption without changing, like a mighty oak in a storm. While durable, a robust system doesn't learn from the experience. This leads to **Resilience**, a more dynamic quality where a system not only survives a shock but recovers its original form, much like a forest regrowing after a fire. Resilience is the ability to bounce back.

These responses are supported by a deeper, internal toolkit. **Plasticity**, especially in the brain, is the remarkable ability to reconfigure internal connections in response to new conditions, allowing us to learn a language or recover from injury [Pascual-Leone2005]. This is often complemented by **Strategic Redundancy**, where multiple functional paths exist for the same task—like having several suppliers for a key component—which provides a crucial buffer against failure. But nature's genius doesn't stop at recovery; it extends to proactive and even opportunistic learning. The engine of this process is **Trial and Error**, the very foundation of natural selection, where what doesn't work informs what might. Sometimes this leads to **Exaptation**, where a trait evolved for one purpose is co-opted for another, like feathers first used for warmth later enabling flight. Other times, it results in pure **Serendipity**, when a mistake leads to a breakthrough, as with the accidental discovery of penicillin [Fleming1929].

This brings us to the most sophisticated strategy: **Antifragility**. Coined by Nassim Taleb, it describes systems that don't just resist stress but actively improve because of it [Taleb2012]. Like an immune system that grows stronger after facing a pathogen, an antifragile system thrives on volatility. Each of these strategies reveals a different relationship with error. Nature's ultimate lesson is not to avoid failure, but to build structures that can absorb it, learn from it, and even use it as an engine for transformation.

Intentional Learning

For millennia, evolution has operated through a seemingly simple yet relentless principle: trial and error. But unlike other organisms, humans have developed an extraordinary ability: the capacity to reflect on our mistakes and learn from them deliberately.

This transition—from a blind and slow evolution to a conscious and directed one—is perfectly embodied in the **scientific method**. This method, which has radically transformed our understanding of the universe, is based on a fundamentally adaptive principle: **falsifiability**. We don't validate hypotheses by their eternal truth, but by their ability to survive rigorous attempts at refutation. When they fail, it is not an interpretation of defeat, but a signal that allows us to reorient the path of knowledge [Popper1959].

In the cognitive realm, this logic is also present. The human mind acts as a **predictive machine**: we anticipate outcomes, form expectations, and when they are not met, we generate a "prediction error." According to research in cognitive neuroscience, this type of error activates specific brain circuits that facilitate the updating and reconfiguration of our mental models [Friston2010]. This neurocognitive model of error-based learning is deeply akin to the scientific model. Just as a scientific theory improves when confronted with unexpected data, our thinking is refined when we are able to tolerate contradiction, reinterpret our beliefs, and generate new hypotheses for action.

This conscious evolution also manifests in organizational and social contexts. The **"Lean Startup"** methodology [Ries2011], for example, is directly inspired by this logic. Building a Minimum Viable Product (MVP), exposing it to the market, and adjusting the design based on detected errors faithfully reproduces the cycle of iterative learning. This structure of constant feedback is, in fact, an industrial formalization of adaptive error.

Therefore, the human ability to learn from error is not an evolutionary exception but a sophistication of a general biological principle. We are the only species that has managed to translate failure into a systematic method for advancement. This conversion of error into a tool is, without a doubt, one of the most extraordinary achievements of consciousness.

Antifragility

The concept of **antifragility**, proposed by Nassim Nicholas Taleb [Taleb2012], represents a fundamental shift in how we understand our relationship with error, stress, and uncertainty. An antifragile system not only withstands disturbances but actively benefits from them. Far from seeking stability, it feeds on the unexpected to become stronger. This principle is particularly relevant in volatile, uncertain, complex, and ambiguous—**VUCA**—environments, where rigid structures collapse and adaptive ones thrive.

In biology, the immune system is a clear example. When exposed to pathogens, it develops immunological memory, becoming better prepared for future attacks. In the psychological realm, certain individuals develop what has been termed **"post-traumatic growth"** [Tedeschi2004]: a capacity to be positively transformed by adverse experiences. These individuals don't just recover; they expand their sense of purpose or appreciation for life. In business, antifragility manifests in models that iterate rapidly, treating failure as just another input for improvement.

The key to developing antifragility is not simply "enduring error" but designing mechanisms that metabolize it as part of growth. This involves mental frameworks that reframe failure, social structures that do not penalize sincere mistakes, and agile feedback systems that transform error into actionable information. Like the tobacco plant that turned a predator into a protector, an antifragile system doesn't just survive its stressors—it uses them to become stronger. This perspective suggests a new way of understanding innovation and evolution: not as error-free conquests, but as developmental trajectories structured by them. In an increasingly uncertain world, it is not enough to be strong or adaptable. It is necessary to be antifragile.

Conclusion

Error, far from being an anomaly, is a structuring function of learning, adaptation, and innovation. The Adaptive Error Thesis proposes that we reframe failure as an evolutionary resource—a necessary and fertile input that can trigger profound transformations. This reframing is not a simple exercise in positive thinking; it is an observation based on scientific evidence. Error activates fundamental processes like brain plasticity, the updating of mental models, and the consolidation of memory. In systemic terms, well-managed error catalyzes higher forms of organization, from resilience to antifragility.

But we don't unlock this adaptive potential automatically. We must create the right conditions: psychological safety, structures that tolerate experimentation, and a cultural willingness to unlearn the stigma of error.

In times of accelerated transformation and radical uncertainty, it is urgent that we abandon the ideal of infallibility and embrace an epistemology of trial. Only by integrating error into our operational logic—at the individual, collective, and systemic levels—can we evolve amidst the chaos. Error, then, is not an obstacle to be eliminated, but a door to be crossed. Whoever passes through it with intelligence and courage does not just survive: they thrive and co-create the future.

THE PERSISTENT PATTERN

To walk along the seashore at dusk is to surrender to a living, natural concert. The air, whistling through the dune grass, carries the faint echo of distant laughter, the sharp cry of a gull spiraling in the vast sky. But beneath this symphony of the ephemeral, like an invisible and tireless conductor, the sea unfolds its melody. It is a persistent song, older than all memory, a planetary breath that never ceases. Its identity is not found in a single moment, but in a pattern that sustains itself across time and space. It is in this paradox—a constant revealed through that which is never the same—that we find a profound echo of our own existence.

Because each of us, too, is an orchestra. A symphony composed of impulses and silences, of tissues that replace themselves, of thoughts that fade and return, of memories that are rewritten. We change, we renew, we forget, and yet, we still recognize ourselves. Something holds steady in that flow: a persistent pattern that, without being static, maintains an internal coherence.

At the heart of our conscious experience lies a continuity that persists beyond cellular renewal and the fleeting nature of thought. This is not an immutable substance, but a dynamic pattern of activity: a functional architecture that, while constantly reconfiguring itself, retains its essential identity.

Contemporary neuroscience suggests the brain operates as a Bayesian prediction system, generating hypotheses about the external world and adjusting its internal models to reduce prediction error [Friston2010]. This process is the foundation of learning, perception, and consciousness. Over time, the patterns of neural activation that support this system stabilize, forming the basis of our identity. They are not fixed anatomical structures, but dynamic configurations emerging from the continuous flow of neural activity, explaining how the "self" remains recognizable amidst constant change.

Literature has long navigated these murky waters. In *One Hundred Years of Solitude*, Gabriel García Márquez portrays the loss of identity through a plague of forgetfulness: when memory is erased, the community unravels, and bonds cease to have meaning [GarciaMarquez1967]. The self-vanishes when it cannot recall its own narrative. At the opposite extreme, Borges gives us *Funes the Memorious ("Funes el memorioso")*, a young man condemned to a perfect memory [Borges1944]. Unable to forget, Funes is a prisoner of his own mental archive; he cannot generalize, abstract, or live freely. These two poles reveal that identity can be sustained neither by forgetting everything nor by remembering everything. It needs to filter, select, prioritize. Our identity rests on a precarious balance between holding on and letting go.

Modern neuroscience reinforces this view. Our memories don't reside in isolated neurons but in distributed networks. Synapses, the points of contact between neurons, are the true carriers of our personal history. Through synaptic plasticity, these connections strengthen with use and weaken with inactivity [Hebb1949]. The brain is not a book, but an evolving melody. Its stability lies not in its components, but in the dynamic architecture of the relationships between them.

More profoundly, science has identified the "molecular engineers" that sustain these networks: proteins like PKMζ, which act as guardians of long-term memory [Sacktor2011]. They are molecules that stabilize synapses and keep the networks running. This system resembles a computer's RAID array: information isn't in one place but is distributed and backed up across multiple nodes. If one part fails, the system can rebuild it. Redundancy protects the pattern. Even the brain's immune cells—the microglia—are no longer seen as mere passive defenders, but as active gardeners of the synaptic landscape, pruning weak connections and consolidating vital ones [Schafer2012].

But this system is not uniform. In childhood, plasticity is extreme. In adulthood, the pattern stabilizes: less flexibility, more robustness. And in old age, the maintenance systems begin to falter. Identity becomes a collective effort of the entire system. This resolves the dilemma. We are not a static object, but a self-organizing process. The continuity of identity is not an illusion, but an emergent property of complex dynamic systems. The "self" is a stable narrative built by multiple biological layers.

However, this system is not infallible. When the maintenance processes fail, the pattern that sustains our identity begins to fracture. Neurodegenerative diseases like Alzheimer's or Parkinson's are not just about cell loss. They represent failures in the brain's ability to sustain its own pattern—errors in the synchronization, cleaning, and reorganization of its networks.

In this context, recent research is broadening its gaze to the gut-brain axis. It has been established that the vast community of microorganisms inhabiting our gut (the micro biome) is not a passive actor but communicates bidirectionally with the brain. An imbalance in this community, known as dysbiosis, is increasingly implicated in the neuroinflammation and progression of neurodegenerative diseases. Disruption of the gut micro biome can compromise the integrity of the intestinal barrier and the blood-brain barrier, allowing microbial metabolites and pro-inflammatory molecules to directly affect brain function [Pathak2024]. The idea is revolutionary and opens new questions: Could a portion of neurodegenerative disorders stem not only from the brain's internal failures but from the complex communication with our gut micro biome? What if the "self" is not only a pattern of neural connections but also the reflection of an interdependent network with the trillions of organisms that cohabit with us?

Furthermore, there are other ways to alter or temporarily suspend the pattern. Drugs can momentarily interrupt the system's coherence. Anesthesia, in particular, represents one of the most intriguing phenomena: one can enter an operating room with an active pattern and leave with the same body, but with no memory of the interval. During that time, the pattern is "switched off" and then restarted [Mashour2013]. This ability to pause consciousness without disintegrating identity raises fascinating questions about the limits of what it means to "be oneself."

The future of neuroscience may require a more integrative view, where the pattern is not only biological or electrical but also ecological. Our being is not a fixed set of bricks, but an energetic dance. There is no immutable essence. There are patterns that renew themselves. There are memories that stabilize. There is a system that maintains, repairs, and reinvents itself... and yet, it remains the same.

PART II

ADAPTATION AND COGNITION

THE INDUCTIVE MACHINE

An Algorithmic Model of Brain Function

In the late 1960s and early 1970s, our understanding of human behavior underwent a significant transformation. The purely reflexive or associationist views of the past were set aside, making way for a vision of the brain as an active, organizing system. This new paradigm presented the brain as an entity capable of simulating, anticipating, and modifying its environment through functional and hierarchical structures.

Among the pioneers of this movement were George Miller, Eugene Galanter, and Karl Pribram. In their seminal work, *Plans and the Structure of Behavior*, they introduced the notion that behavior is mediated by "plans." These plans could be broken down into fundamental operational units they called TOTE (Test–Operate–Test–Exit) [Miller1960]. The TOTE model replaced the traditional stimulus-response (S-R) framework by demonstrating that human behavior results from an internal process of comparison, action, evaluation, and conclusion, much like the logic of a control program. According to Miller and his colleagues, an organism assesses its current situation against a goal, acts to reduce any discrepancy, re-evaluates, and exits the process when the difference is eliminated. This framework profoundly influenced cognitive psychology and the design of artificial intelligence systems, but it assumed a functional pre-programming that didn't fully address how new heuristic structures are generated.

At almost the same time, Peter Putnam developed a model of the brain that went beyond feedback control, proposing a view of it as a "self-organizing inductive machine" [Putnam1966]. His model, based on the "resolution of internal contradictions" (which he termed "X"), allows the brain to generate meaning and behavior. Unlike the TOTE approach, the inductive model presents a system where the cognitive structures themselves—such as criteria, acts, search, and inhibitions—are dynamically reconstructed as contradictions emerge. For Putnam, human thought is not modeled as a flow directed by pre-established plans, but as an emergent system in which contradictions between acts and contexts trigger processes of search, synaptic modification, and the generation of new heuristics.

Putnam's algorithmic model of brain function describes an iterative functional sequence where learning and behavior emerge from the active confrontation with these internal contradictions. Putnam's fundamental premises include the possibility of understanding the brain through reverse engineering methods, conceiving of cognitive life as a "single-player game of mathematical game theory." The brain operates by collecting correlations through all-or-nothing "acts," which are the primary functional units. A contradiction ("X") arises when a successful act is over-generalized, and its resolution triggers serial (SE) or random (RS) search processes. An internal random search is activated when the serial search is undermined, generating alternative acts until a stable state is achieved, which then becomes new knowledge.

In this system, the "rules" of the Inductive Model manifest through "functional protocols," which are recurring sequences of adaptive behavior. These protocols are not fixed algorithms but structures that coordinate the application of multiple cognitive rules. Several key protocols are identified:

Contradiction Resolution Protocol (CRP): Activated by prediction errors or adaptive failures. It detects an "X" (conflict), suspends the automatic response, triggers an internal random search, explores alternatives, and consolidates the new solution.

Heuristic Generation Protocol (HGP): Activated by the repetition of successes. It detects the possibility of generalizing a successful act, evaluates its contextual limits, introduces variations, and stabilizes valid extensions as new heuristics.

Creative Exploration Protocol (CEP): Activated by periods of stability. It generates self-induced "X"s (curiosity, play), triggers an internal random search without external stimulus, builds mental scenarios, and tests them.

Social Simulation-Coordination Protocol (SSCP): Activated by interaction with others. It identifies social "X"s (interpersonal dissonance), simulates the other's perspective (Theory of Mind) [Premack1978], adjusts one's own acts based on the model of the other, and establishes shared rules.

These protocols do not operate in isolation but are intertwined. Putnam's model, therefore, not only describes how a brain thinks but how a universal cognitive system operates, where contradiction is not a failure but the starting point for reorganizing reality.

The Fruit Fly Larva as a Minimal Inductive Machine

During a casual observation, the adaptive behavior of a larva (fruit fly maggot) was witnessed as it left a fallen mango. The larva began to move, guided by rudimentary sensory patterns. When a path was drawn on the ground with a moistened finger, the maggot followed this trajectory, guided by its sensitivity to moisture.

The maggot's system operated on a simple heuristic: "move toward areas of familiar moisture." This rule was sufficient until an experimental contradiction was introduced: a small puddle of water that offered greater moisture. Upon entering, the maggot showed signs of agitation. It then retreated and avoided entering similar puddles again. It had modified its original heuristic: extreme moisture had become a condition for rejection.

This event reveals the operation of the **Contradiction Resolution Protocol (CRP):** an X is presented (excess moisture), an exploration occurs (RS), an adjusted response is developed, and the new rule is stabilized. Subsequently, at the edge of another puddle, the maggot moves around the obstacle, demonstrating the emergence of a new functional heuristic via the **Heuristic Generation Protocol (HGP)**. This simple organism proves to be a biological inductive machine.

A Newborn in Search of the Nipple

During the first days of life, a human newborn displays a basic yet powerful behavior: when brought close to its mother's breast, it begins a rhythmic, erratic movement. This activity is guided by innate impulses but lacks precise direction.

The moment its cheek or lips brush against the nipple, a different sequence is activated: the mouth opens, the head turns, and the sucking reflex occurs. This transition marks the resolution of a sensory contradiction (X): the conflict between the feeding drive and the absence of a clear stimulus.

This dynamic precisely reflects the **Contradiction Resolution Protocol (CRP)**:

- The sucking impulse is present but finds no object → **X**.
- An external random sensory search (RS) is activated.
- Detection of the nipple acts as a marker → activates the organized response.
- The sucking pattern is consolidated, and with repetition, the initial search shortens.

This example shows how a minimal nervous system resolves contradictions from primary impulses and consolidates new heuristics. The inductive machine does not need language to function; interaction, error, and reorganization are sufficient.

Systems Where the Model Applies

One of the most notable features of the Inductive Model is its functional universality. It can be applied to viruses, cells, plants, animals, humans, and organizations. This makes it a unifying meta-model, where all these systems operate under the pressure of contradiction, explore through search, and consolidate solutions through structural stabilization.

Level	Example	Form of "X" Resolution
Virus	Host cell entry strategy	Reorganization by selecting functional genetic pathways
Cells	Response to contradictory chemical signals	Biochemical search and selective receptor activation
Plants	Adjustment of tropisms to opposing conditions	Reorientation based on light/gravity/moisture correlations
Animals	Adaptive behavior in the face of ambiguity	Activation of internal RS and modification of acts
Humans	Complex decision-making	Resolution of cognitive contradiction, symbolization, learning
Organizations	Institutional crisis or cultural change	Generation of new protocols and cooperative structures

In conclusion, Putnam's model offers a foundation for understanding brain function as a heuristic, dynamic, and adaptive system. Cognitive heuristics are not fixed rules but functional solutions that emerge in protocols activated by contradictions. This architecture can be extended to artificial systems and collective decision-making processes, offering a unified conceptual basis for the study of cognition, learning, and adaptive behavior.

Each protocol is not an isolated function but a dynamic set of adaptive operations that recruit different cognitive heuristics as needed. These heuristics, in turn, can be understood as implicit processing rules that have been evolutionarily selected for their effectiveness in solving certain types of problems. Some are general and are activated across multiple domains, while others emerge as specific sub-rules or corollaries under certain conditions.

The following table summarizes the general organization of these protocols and their associated heuristics:

Protocol	Heuristics
Internal Contradiction Resolution: Detects prediction errors, conflicts between acts and context, or emotional dissonance. Suspends automatic responses, activates an internal search, and reorganizes behavior or cognition to restore adaptive coherence.	**Primary:** Prediction and Error, Memory and Cognitive Readjustment, Cognitive Resilience, Defense Mechanisms, Energy Economy. **Sub-rules:** Biases such as negativity bias, emotional memory, rumination, cognitive closure.
New Rule Generation: Generalizes successful acts and allows for their flexible application in new contexts. Stabilizes useful solutions, reduces future adaptive load, and builds functional automatisms.	**Primary:** Repetition with Variation, Synaptic Plasticity, Contextual Transfer, Pattern Recognition, Motivation by Incompleteness. **Sub-rules:** Preference for simplicity, functional automation, consolidation of useful habits.
Creative Exploration: Activated in conditions of prolonged stability or in environments rich with possibilities. Allows for the simulation of new configurations without direct risk, favoring imagination, novelty, and divergent thinking.	**Primary:** Play and Simulation, Divergent Thinking, Heuristic Curiosity, Active Imagination **Sub-rules:** Anticipation, emotional regulation, novelty seeking.
Social Coordination and Simulation: Allows for representing the intentions of others, establishing symbolic communication, and stabilizing shared rules. Optimizes group interaction and the transmission of learning.	**Primary:** Theory of Mind, Language and Symbolization, Social Imitation, Collective Memory. **Sub-rules:** Tactical empathy, cultural learning, internalized social norms.

BIOLOGICAL CONTINUITY

When I was in elementary school, my science teacher told us about a theory that stuck with me. He described a hypothesis that, at some point in our evolution, all species shared a behavior very similar to that of plants. In the beginning, life on Earth was dominated by simple, photosynthetic organisms that existed in a kind of symbiosis with their environment.

I remember my astonishment, the question that sprang from my childhood mind: "Wait, we were all *plants*!?" That exclamation, loaded with a mix of disbelief and fascination, captured the essence of an idea that today, all these years later, still resonates with a solid foundation in the principles of evolutionary biology and comparative physiology.

It was in this context of early life that the story took an unexpected turn. A bacterium, likely something similar to the mitochondria we find in our cells today, integrated itself within other organisms. It wasn't a hostile invasion, but a profound symbiosis. This bacterium brought something crucial to the table: the ability to generate energy much more efficiently. Over time, the organisms that hosted it began to evolve in more complex ways.

The Last Universal Common Ancestor (LUCA), while not photosynthetic itself, possessed primitive metabolic pathways that laid the groundwork for the complex bioenergetic processes that would follow [Weiss2016]. Among its early descendants, cyanobacteria stand out. These photosynthetic prokaryotes not only pioneered the oxygenation of the atmosphere but also developed the first biological timing mechanisms, synchronizing themselves with the cycles of sunlight. This ability to anticipate and respond to environmental variations is considered a fundamental precursor to the modern circadian clock [Dvornyk2003].

The evolution of biological complexity took a quantum leap through endosymbiosis. Cyanobacteria were incorporated by ancestral eukaryotic cells, giving rise to chloroplasts. In parallel, the integration of aerobic bacteria evolved to become mitochondria [Margulis1970]. This process was not a mere annexation but a deep fusion where the ancestral bacterium contributed significantly more efficient energy production—a cornerstone of growing biological complexity.

Our physiology remains intrinsically dependent on sunlight for processes like vitamin D synthesis and the regulation of circadian rhythms. Exposure to ultraviolet B (UVB) radiation on the skin triggers the synthesis of cholecalciferol, a precursor to vitamin D, which is crucial for calcium homeostasis and immune function [Holick2007]. In the hypothalamus, the suprachiasmatic nucleus (SCN) acts as the master circadian pacemaker, receiving photic information directly from the retina. This nucleus coordinates functions such as hormone release and sleep-wake cycles [Moore2002].

Virtually all human cells possess intrinsic circadian oscillators, a parallel to the leaf movements in plant organisms. At a molecular level, this kinship is evident in the conservation of photoreceptors. Proteins like cryptochromes (CRY), present in both plants and animals, mediate responses to blue light and are essential components of the circadian clock [Cashmore1999].

The ancient mitochondrial symbiosis has revealed a deeper dimension of cell biology. Mitochondria, once independent bacteria, persist as vital organelles. Beyond their role in energy production, recent discoveries suggest that the human body operates as an interconnected network, where every cell actively participates in information processing. Concepts previously attributed exclusively to the brain—such as decision-making, adaptation, and learning—are now being observed at the cellular level [Levin2021].

These capabilities are manifestations of Putnam's Inductive Machine within each biological unit.

This "cellular intelligence" transcends mere metaphor. Cells communicate through a complex network of chemical, mechanical, and electrical signals. The nineteenth-century conception of the cell as a passive unit has been overturned. Eric Kandel's work showed that neurons can store information through synaptic plasticity [Kandel2001]. Subsequent research has confirmed that immune, cardiac, and even skin cells can "remember" and modify their behavior. A recent study showed that non-neuronal cells can also activate genes related to cellular memory, including the transcription factor CREB [Huang2024].

Today we know that cells not only communicate chemically but also use bioelectric patterns to coordinate collective actions [Levin2021], a silent language expressed even in tissues without neurons. This allows them to develop structures and regenerate parts, as if a form of distributed intelligence lives within them.

All of this leads us to a deeper reflection: the brain is not a solitary director. It is a node in a complex network that runs through the entire body. What we call the mind appears to be an emergent property of the organism as a whole. Even without chlorophyll, we share with plants a sensitivity to light inscribed in our physiology. The story of that ancestral bacterium that chose to collaborate is still alive within us. Not as a memory, but as a functional structure. Understanding this reconfigures our idea of what we are: a profoundly integrated, intelligent, and constantly evolving system.

That is why, when observing an oak tree at sunset, one doesn't just contemplate a botanical phenomenon. One perceives a shared resonance, a biological echo. Because even though we don't photosynthesize, we too are children of the sun.

Author's Note:

It's possible I wasn't really paying much attention in class when my science teacher talked about this theory. Maybe I confused concepts or filled in the gaps with my imagination. Or maybe—and I like to think this—my childhood mind connected loose ideas in such an original way that it ended up recreating, almost by accident, a real scientific theory: Lynn Margulis's endosymbiotic theory, proposed in 1967.

THE DANCE OF ORDER AND CHAOS

The Second Law of Thermodynamics teaches us that in an isolated system, entropy—a fancy word for disorder—tends to increase over time. This might lead you to expect a universe marching relentlessly toward a state of maximum randomness. Yet, in the realm of complex systems, we often see phenomena that, at first glance, seem to fly in the face of this trend. A complex system, unlike one that's merely complicated, is defined by non-linear relationships between its parts, a dependence on its own history, fuzzy boundaries, and feedback loops that can either amplify or dampen effects [Buzsaki2006]. This is where Chaos Theory comes in. Far from the pop-culture idea that "everything descends into chaos," Chaos Theory actually studies deterministic systems that, despite following precise rules, exhibit an extreme sensitivity to their starting points, which can generate unpredictable behaviors and surprisingly complex patterns [Strogatz2014].

This "unpredictability" doesn't stem from inherent randomness, but from an exponential sensitivity to initial conditions. The tiniest, most minuscule variation in a system's starting point can lead to radically different future paths. In practice, our inability to measure initial conditions with infinite precision makes long-term prediction impossible—not because the system isn't following deterministic rules, but because of the inherent limits of our observation and calculation.

It is in this context of apparent randomness that we witness moments of astonishing, spontaneous order: metronomes syncing up, moons orbiting with flawless precision, fireflies flashing in unison, and even the steady beat of your own heart [Strogatz2003]. These examples show how, even in systems

75

governed by the thermodynamic push toward disorder, phenomena of self-organization and synchronization can emerge. What principles underlie this order that defies the natural inclination toward chaos?

Before modern developments in complex systems theory, thinkers like Peter Putnam were already proposing functional models of the nervous system. Their work sought to understand how the brain establishes relative dominance between competing actions and how sensory factors interconnect these loops [Putnam1966]. They built on Hebb's notion of "resonant neural loops" [Hebb1949] as representatives of motor acts. Although their approach was centered on the conditioned reflex as the basic building block and they acknowledged inhibition as a mechanism for modulating neural activation, their perspective differed from more contemporary views that emphasize inhibition's fundamental role in stability and the generation of large-scale complex rhythms. Their model, however, already anticipated the idea of self-facilitating cyclical patterns and the importance of the brain's internal organization for behavior, laying the groundwork for understanding the emergence of order in complex biological systems.

The Sympathy of Pendulums

In 1656, the celebrated Dutch physicist Christiaan Huygens created the first functional pendulum clock. His goal was to help sailors determine their position in the world. While latitude could be measured by observing the sun or stars, longitude required knowing the exact time at a fixed point, like a home port. The clocks of the era were useless, losing up to 15 minutes a day. In contrast, Huygens's pendulum clocks were accurate to within 10 to 15 seconds per day.

Huygens's plan was to attach his clocks to a heavy weight on the ship so they wouldn't be affected by the tides. He needed two clocks, just in case one stopped or was damaged. It was while testing this setup, sick at home in February 1665, that he made a curious discovery. Two of his clocks were hanging from a single wooden beam propped up on two chairs. As he watched the pendulums swing for hours, he noticed that after about half an hour, they would spontaneously sync up. When one clock swung one way, the other swung the opposite way; when one went "tick," the other went "tock."

He tried to break this sync by knocking them out of phase, but once again, after about thirty minutes, they returned to their synchronized dance. Huygens initially chalked this "odd sympathy" up to air currents between the pendulums, so he placed a board between them. Yet, the clocks continued to sync. It wasn't the air currents. When he moved the clocks apart, the synchronization vanished and their times drifted, but when he brought them close again, the sync returned.

Then it dawned on him: the clocks were syncing up because they were hanging from the same wooden beam. The beam was transmitting tiny mechanical vibrations from one clock to the other, coupling the two oscillators. He was the first to observe this kind of spontaneous synchronization in inanimate objects [Strogatz2003]. Although he qualitatively described what he saw, it was only a few decades ago that scientists began to build a rigorous theory of synchronization.

The Kuramoto Model

Picture several metronomes placed on a board that can wobble slightly. If you start them at different times, at first they'll each keep their own beat, creating a chaotic racket. After a little while, however, something amazing happens: they begin to synchronize until they all swing in perfect unison. This magical phenomenon occurs even if the metronomes don't have the exact same natural speed.

77

The key is the board that connects them. When a pendulum swings to the left, it pushes the board a tiny bit to the right, and vice versa. This small vibration is transmitted to all the other metronomes. If one metronome is out of step, the motion of the board, driven by the majority, gives it a little nudge with each cycle, subtly adjusting its rhythm until it falls in line with the others. It's an indirect conversation that leads to collective order.

To understand this mathematically, we can visualize each metronome's cycle as a point moving around a circle. This concept is called the "phase" of the oscillator. Imagine the rightmost point of the circle is the start of the cycle (0 degrees) and the leftmost point is the halfway mark (180 degrees). As the metronome's pendulum swings back and forth, its corresponding point makes a full lap around the circle. A faster metronome simply means its point spins at a higher speed.

When the metronomes are out of sync, their points are scattered around the circle, each moving at its own pace. When they finally synchronize, all the points cluster together and circle around as a single swarm.

This idea is the basis of the Kuramoto Model [Kuramoto1984]. This mathematical model describes how the speed of each oscillator (each point on the circle) depends not only on its natural frequency (its individual rhythm) but also on the influence of all the others. This influence is known as the "coupling strength." If this force is strong enough, it can overcome individual differences and "drag" all the oscillators into a common, synchronized rhythm.

A good analogy is a group of friends jogging on a track [Strogatz2003]. Each person has their own natural pace. If a faster runner encourages a slower one, and the slower one tries to keep pace, they become "coupled." If this coupling is strong, they'll adjust their speeds to run together. But if the speed difference is too great or there's no motivation to sync up (weak coupling), they'll end up doing their own thing.

Phase Transitions

The fireflies of Southeast Asia synchronize their flashes even though each has its own internal frequency. They couple with each other so distinctly that hundreds, even thousands, can flash together in the same fraction of a second. There's a great simulation of this by Nicky Case. You start with individual fireflies flashing at their own pace, and then you can turn on the interaction between them. In the Kuramoto model, this would mean each firefly affects all the others. But in this simulation, a firefly is only affected by its neighbors: if it sees a nearby flash, it nudges its internal clock forward slightly, flashing a little sooner than it normally would. What's amazing is that even though the interactions are small and local, after a while, you see waves traveling through all the fireflies, and eventually, they all flash at the same time.

You might think that as you gradually increase the coupling, you'd get a system that's more and more in sync. But that's not what happens. Think about how water freezes: it doesn't get progressively icier. It's liquid, and then, at a critical temperature, the molecules abruptly shift state and become solid. This is a temporal version of that same idea. It's as if the fireflies' phases "lock in" once they cross a certain coupling threshold. At that point, this kind of "crystallization in time" is the phenomenon we call synchronization [Winfree1980]. Phase transitions are abrupt, qualitative shifts in a system's behavior when a control parameter (in this case, coupling strength) crosses a critical value, similar to how a material goes from liquid to solid or from paramagnetic to ferromagnetic.

The Universality of Synchronization

One of the most fascinating things about synchronization is its universality. It happens on every scale of nature, from the subatomic to the cosmic, using every channel of communication nature has created: gravitational, electrical, chemical, mechanical interactions... Any way that two objects can influence each other is used by nature for synchronization [Strogatz2003].

The cosmos is a vast stage where countless celestial bodies perform an intricate ballet, governed by the unwavering laws of gravity. Among the most fascinating phenomena in this cosmic dance is synchronization, a process through which the rotational and orbital rhythms of celestial objects adjust and harmonize. A prime example of this gravitational harmony is tidal locking, an effect that is ubiquitous in our own solar system.

Let's take the most familiar case: our Moon. The fact that it always shows us the same face isn't a coincidence; it's the direct result of tidal locking. The Moon rotates on its axis exactly once for every orbit it makes around the Earth. This phenomenon isn't unique to our Moon; it has been observed in 34 moons in our solar system that exhibit this particular relationship with their host planets.

The mechanism behind tidal locking is a subtle but powerful dance of gravitational forces. A moon, initially with its own rotational frequency, experiences a slightly stronger gravitational pull on the side closer to the planet. This difference in gravitational force deforms the moon, stretching it slightly into an elliptical shape. As the moon continues its orbit and rotation, these tidal "bulges" are constantly pulled out of alignment with the planet. The planet's gravitational force, acting on these bulges, constantly tugs them back into alignment, creating a torque. This torque acts as a brake on the moon's rotational speed, gradually slowing it down until it becomes locked with the planet. It's crucial to note that if, for some reason, the moon were rotating too slowly initially, this same mechanism could speed it up until it reached the locked state. Gravity, in this sense, seeks an equilibrium, pushing or pulling until synchronization is established.

Tidal locking is just one of many manifestations of cosmic synchronization. There are other types of synchronization phenomena, equally beautiful and complex, that manifest in gravitational interactions. An outstanding example is found in the three moons closest to Jupiter: Io, Europa, and Ganymede. These moons are not only tidally locked with Jupiter but also engage in a fascinating 1:2:4 orbital resonance with each other.

This resonance means that for every one time Ganymede, the outermost of this trio, orbits Jupiter, Europa completes two orbits and Io, the innermost moon, completes four. This precise mathematical relationship is the result of a delicate gravitational balance that has been maintained for billions of years. The gravitational forces the moons exert on each other and on Jupiter have pushed them into this stable configuration, where their orbital periods are related by simple integer ratios. Orbital resonances have profound implications for the stability and evolution of planetary systems, influencing the shape of orbits and the distribution of energy within the system.

These examples, from the tidal locking of our Moon to the intricate orbital resonances of the Jovian moons, offer us a captivating glimpse into how gravity orchestrates a cosmic ballet of synchronicities. They are a testament to the elegance and precision inherent in the fundamental laws that govern our universe, revealing an underlying order in the apparent vastness and complexity of space.

In the 1950s, Russian scientists, including Boris Belousov and later Anatol Zhabotinsky, sought a chemical reaction that would repeatedly alternate in color, much like a pendulum swings back and forth. Although thermodynamics dictates that closed systems tend to increase entropy and reach equilibrium, there is no principle preventing a gradual oscillation as the system approaches that equilibrium. This is precisely what Belousov and Zhabotinsky discovered [Zhabotinsky1991].

Even more spectacular is the BZ reaction when performed without stirring, such as in a Petri dish. In this setup, one can observe spiraling waves of color or expanding circular patterns. It's crucial to understand that the liquid itself is not moving; these are not like ripples in a pond. Instead, what is moving are the chemical concentrations. You can see these blue "chemical waves" propagating and moving at a constant speed, or watch spirals that continuously grow and rotate.

The BZ reaction beautifully illustrates that even complex systems governed by thermodynamic principles can exhibit dynamic, organized behavior as they move toward equilibrium. What's most striking about this is that the same phenomenon is seen in the heart. You can see spiral waves of electrical excitation in a heart that look exactly like the spiral waves of the chemical oscillations, like the chemical waves in the BZ reaction.

Synchronization in the Brain

The brain is the quintessential example of a complex system, where synchronization and desynchronization play crucial roles in its function.

Contrary to intuition, neural inhibition is just as vital as excitation for generating rhythmic patterns and enabling complexity to emerge. Purely excitatory neural networks are inherently unstable. The brain's "tensegrity harmony"—its structural and dynamic stability—is achieved through a delicate balance between excitation and inhibition. Inhibition introduces non-linear effects that allow for the spatiotemporal segregation of principal cells and "winner-take-all" mechanisms, which are essential for the selection and differentiation of neural assemblies. This excitatory/inhibitory (E/I) balance is fundamental to the robustness of neural networks and their ability to respond to subtle stimuli [Buzsaki2006].

Brain connectivity is not random; it follows principles of "small-world" and "scale-free" networks [Buzsaki2006]. "Small-world" networks combine dense local clustering with a few long-range connections that drastically shorten the average path between any two neurons. This allows for efficient large-scale communication with minimal wiring. "Scale-free" networks, on the other hand, are characterized by a connectivity distribution where a few "hubs" (highly connected nodes, like the locus coeruleus in the brain) have a disproportionate influence, which grants the network robustness against random failures, though it makes it vulnerable to targeted attacks on these hubs [Buzsaki2006].

Brain activity is largely self-generated or spontaneous. External sensory inputs are not simply processed passively; rather, they cause a slight deviation or modulation of this robust internal program. The brain's intrinsic "state," influenced by its history and spontaneous activity, is a crucial determinant of how stimuli are interpreted and responses are generated. The variability in neural responses, often dismissed as "noise," may actually be a functional manifestation of this spontaneous activity and the brain's contextual interpretation. Phenomena like stochastic resonance demonstrate how noise, paradoxically, can enhance the detection and transmission of weak signals in non-linear systems [Buzsaki2006].

A fascinating example of brain synchronization, especially during rest, is the Default Mode Network (DMN). This network of brain regions, including areas like the medial prefrontal cortex, posterior cingulate cortex, and inferior parietal lobule, becomes consistently active when an individual is not focused on external tasks and the brain is in a state of wakeful rest [Raichle2015]. Activity within the DMN is highly synchronized and is associated with internal cognitive processes such as introspection, future planning, episodic memory retrieval, and the consolidation of information patterns during rest. Its functional coherence underscores how the brain, even in the absence of direct external stimuli, maintains a state of internal organization that is fundamental for consolidating experience and preparing for future interactions.

Finally, there is a fundamental tension: complexity and synchrony compete with each other [Buzsaki2006]. Excessive synchronization in the brain can reduce its capacity to process complex and differentiated information, leading to pathological states like epileptic seizures. Therefore, a dynamic balance between synchronization and desynchronization is essential for the richness and efficiency of the brain's information processing [Singer1999].

In the realm of general biology, we observe how organisms' circadian rhythms intrinsically synchronize with the day-night cycle. This synchronization, mediated primarily by light through structures like the suprachiasmatic nucleus in mammals [Moore2002], is a fundamental mechanism that regulates patterns of sleep, feeding, body temperature, and hormonal activity, and is crucial for the health and well-being of living things. Similarly, in some animal and plant species, the reproductive cycles of entire populations can be synchronized. This collective phenomenon, often influenced by environmental cues or pheromones, increases the chances of reproductive success and species survival by maximizing opportunities for mating and resource availability for offspring, as seen in the mass spawning of corals or the synchronous flowering of some bamboo species.

From the perspective of physics and technology, synchronization is a key operating principle. A classic example is the laser, whose functionality is based on the coherent and synchronized emission of billions of photons. All these photons vibrate in phase and with the same direction and frequency, resulting in a highly directional and monochromatic beam of light. In the field of electronic circuits, the synchronization of oscillators is fundamental to the operation of a vast range of devices, from radios and televisions to computers and communication systems. Ensuring that signals are transmitted and received correctly depends on the precision with which oscillating components maintain their phase. A critical example in engineering is the synchronization of generators in a power grid, where all must operate in perfect phase and frequency synchrony to prevent massive power outages [Strogatz2003]. At a more fundamental level, quantum phenomena like superconductivity and super fluidity are astonishing examples of how particles (electrons in superconductors or helium atoms in superfluids) can behave collectively and synchronously at extremely low temperatures, exhibiting unique macroscopic properties such as the absence of electrical resistance or viscosity. These states of matter demonstrate that synchronization can emerge even at microscopic scales, with profound macroscopic implications. Even in mechanical systems, synchronization is vital; consider the rhythmic coordination of an insect's legs as it walks or the precise synchronization of an airplane's engines, where coherence in motion is essential for efficiency and stability.

In science, we practice reductionism. Our science classes teach us that the way to solve a problem is to break it down into its parts and analyze each one. And that has been phenomenally successful for every branch of science. But the great challenge for science today is what happens when you try to step back and put all the pieces together to understand the whole. That's the domain of complex systems. It's why we don't fully understand the immune system, or consciousness, or the economy. The apparent emergence of properties of the whole that cannot be simply deduced from the properties of its parts constitutes a central challenge in contemporary research: understanding how interactions at the component level give rise to collective, functional phenomena at the system level.

Despite advances in understanding synchronization, significant challenges and open questions remain. For example, in neuroscience, the "binding problem"—how the brain integrates diverse and distributed sensory information into a unified, coherent perception—remains an active area of research, with neural synchronization as a central hypothesis [Singer1999]. Likewise, the transition from synchronized neural activity to conscious experience remains one of the great unanswered questions in science [Mashour2013]. These challenges underscore that while we have unraveled many principles of emergent order, the inherent complexity of natural systems continues to offer vast frontiers for scientific exploration.

THE NATURE OF TIME

The nature of time, a fundamental construct in both human experience and physics, is intrinsically tied to the brain's capabilities and limitations. This organ, arguably the most sophisticated device known in the universe with its roughly one hundred billion neurons, was not, however, "designed" for the express purpose of understanding the nature of time or even for analyzing itself [Buonomano2017]. Consequently, the intuitions and theories we've developed about time reveal as much about the brain's architecture and inherent constraints as they do about the essence of time itself. It has been noted that grasping time is more challenging for the human intellect than grasping space, as animal brains are generally better equipped for navigating, perceiving, representing, and understanding spatial dimensions. A leading hypothesis suggests that the human ability to conceptualize time was "hijacked" from neural circuits that evolved primarily for understanding space [Buonomano2017]. It's crucial to emphasize that the brain does not directly "perceive" or "interact" with the external world; all information is received through patterns of action potentials generated by our sensory organs, which are then interpreted and synthesized into an internal representation of reality [Koch2004]. This inherent limitation of the brain, which is evident even in seemingly simple tasks like numerical calculations, highlights how its architecture shapes our ability to tackle questions it didn't evolve to answer—including the nature of time itself [Buonomano2017]. Our understanding of time is further complicated when we distinguish between natural time (derived from cosmological phenomena like Earth's rotation), clock time (a human invention for precise measurement), and subjective time (our internal, plastic experience of its passage).

The empirical observation underlying this concept is the inherent ability of cells to generate time. From single-celled organisms like cyanobacteria (*Synechococcus elongatus*), which regulate their photosynthetic metabolism using protein-based circadian oscillators (KaiA, KaiB, KaiC) [Uetake2001], to the synchronized opening of stomata in plants or the programmed nighttime sporulation in fungi like *Neurospora crassa*, cells manage temporal events through endogenous clocks [Dunlap1999]. These biochemical systems mark the passage of time without external reference, constituting an emergent property of their internal rhythms, synaptic delays, and metabolic oscillations. Each cell, in this sense, can be considered a living "temporal particle," whose coordination gives rise to a higher-order temporality in tissues, organs, and, ultimately, the mind. The brain, as the pinnacle of cellular organization, thus becomes the most sophisticated laboratory for this temporal fabrication. The evolution of these cellular circadian clocks, as the "escape from light" hypothesis suggests, may have been driven by the need to coordinate vital functions, like cell division, with cycles of light and darkness to avoid UV radiation damage [Pittendrigh1993]. The synchronization of these internal clocks with external environmental cues, known as *zeitgebers*, is fundamental to biological adaptation.

The "escape from light" hypothesis posits that the first single-celled organisms developed internal oscillators to anticipate and adapt to cyclical environmental changes, specifically the day-night cycle [Pittendrigh1993]. This early adaptation would have conferred a significant selective advantage, allowing cells to optimize metabolic processes like photosynthesis during the day and DNA repair or cell division at night, minimizing exposure to harmful ultraviolet radiation. Thus, the ability to "measure" time internally, without relying on the direct perception of light, became a fundamental survival mechanism, laying the groundwork for the temporal complexity seen in multicellular organisms and, eventually, in the brain.

Although the brain doesn't grant us the ability to physically move through time, it can be considered a "time machine" by virtue of four interrelated functions essential for survival and adaptation [Buonomano2017]. First, it operates as a prediction and anticipation machine, using information from the past to forecast future events—a crucial function for survival and reproduction [Bubic2010], [Hohwy2013]. This process involves making automatic, unconscious short-term predictions (like adjusting your movement to catch a ball) and storing vast amounts of past information, often with implicit time stamps, allowing for the review of life episodes organized on a timeline. Second, it serves as a time-measuring machine, performing a wide range of computations for recognizing and generating temporal patterns, from musical rhythms to regulating circadian sleep-wake cycles [Ivry2004]. Third, the brain creates the subjective sense of time. Unlike sensory modalities such as vision or hearing, there is no specific sensory organ for time. The conscious feeling of time's passage and the duration of events is, therefore, a mental construction, constituting one of the most fundamental and universally shared experiences, though it is susceptible to numerous illusions [Wittmann2013]. Finally, the brain enables mental time travel. Humans possess the ability to cognitively project themselves into the past (episodic memory) and the future (prospective simulation) [Schacter2007]. This capacity, considered a key factor in our species' advancement [Suddendorf2007], makes it possible to recall past events, simulate various future scenarios, and plan present actions to shape the future, which has been fundamental to the development of culture and technology.

The brain's processing of time is articulated through intricate and distributed mechanisms. Temporal relationships between events are essential cues the brain uses to make sense of the world. The order and interval of events shape neural circuits, and the rules of causality are encoded at the synaptic level [Hebb1949]. Temporal contiguity is, in fact, fundamental to learning processes like classical conditioning (e.g., Pavlov's dog) [Rescorla1988] and object recognition (e.g., the size-invariance problem in a baby's perception). The latter refers to the brain's ability to recognize that an object maintains its real size, regardless of its distance or the size it projects on the retina. To a baby, an approaching object appears to get bigger, but the brain quickly learns to interpret this change in retinal size as movement in depth, not as the object itself growing. The brain assumes that different sensory patterns occurring in close temporal succession come from the same object, allowing it to build a coherent representation of the world [Buonomano2017].

As for timing mechanisms, the brain operates on a **"multiple clock principle,"** meaning there are various mechanisms for tracking time on different scales, from milliseconds to years, in contrast to the single clock of artificial devices. There is no singular "master clock" governing timing at all scales; rather, most neural circuits have an intrinsic ability to measure time when the situation demands it, this being a fundamental property of neurons [Mauk2004]. Short-term synaptic plasticity (STDP), for example, contributes significantly to the brain's ability to measure time on the order of hundreds of milliseconds by strengthening or weakening synapses based on the temporal sequence of neural activity, acting as a neural detector of cause and effect [Dan2006]. One of the ways the brain encodes time is through **"population clocks,"** where each moment is represented by an active subpopulation of neurons. Furthermore, the brain leverages the intrinsic dynamics of its neural networks to correlate internal states with changes in the external environment. Simulations have shown that recurrent neural networks can generate complex and reproducible patterns of activity, essential for motor control and timing. In these networks, information manifests as an emergent property, distributed across synapses and neurons, where the functional whole is greater than the mere sum of its individual parts [Rolls2010]. The state of a neural network, including its "hidden state" (time-varying synaptic properties), influences its response to future stimuli, allowing the network's "past experience" to be stored in its current state, much like how ripples in a pond reflect the sequence of raindrops [Buonomano2017].

The subjective perception of time is a mental construct susceptible to temporal illusions and distortions, influenced by external and internal factors like cognitive load, which can result in the overestimation or underestimation of event durations [Eagleman2008], [Droit-Volet2016]. It has been documented that life-threatening situations can radically alter the subjective perception of time, causing an apparent slowdown (the **"slow-motion effect"**). Various hypotheses attempt to explain this phenomenon: the "overclocking" hypothesis suggests accelerated brain processing; "hyper-memory" posits a greater retention of details in the post-event recall; and the "meta-illusion" proposes that the perception of time's speed is an arbitrary setting imposed by consciousness, making the slowdown an "illusion of an illusion" [Eagleman2008]. Phenomena like the **stopped-clock illusion**, where the first second on an analog clock seems to last longer, illustrate how our own actions (like eye movements) can distort temporal perception. Psychoactive substances, such as cannabis or dopaminergic drugs, can also drastically influence the sense of time, often interpreted as an alteration in the speed of a hypothetical internal clock, although results can vary depending on the task and the interval studied. Additionally, the brain has the ability to "recalibrate" the perception of events, integrating visual and auditory signals that arrive with different latencies into a unified experience (e.g., thunder and lightning).

The brain's ability to subjectively compress and dilate time is not just a distortion but a functional feature. The ability to mentally replay a song at different speeds or the observation of accelerated neural "replays" of spatial trajectories in the hippocampus of rats during sleep or rest [Buzsaki2006], suggest that the brain can process and generate temporal patterns at variable speeds, which could contribute to memory formation and future action planning.

Several brain areas have been identified as relevant for time processing. The **Suprachiasmatic Nucleus (SCN)**, located in the hypothalamus, acts as the brain's "master circadian clock," regulating the body's daily rhythms (sleep-wake, body temperature) and synchronizing other cells [Dunlap1999]. The **prefrontal cortex**, notably expanded in primates and humans, is crucial for higher cognitive functions like long-term planning, delaying gratification, and mental time travel. The **hippocampus** contains "place cells" and neurons that encode time and distance, making it essential for episodic memory, which, in turn, is fundamental for mental time travel. The **basal ganglia** and the **supplementary motor area** are involved in discriminating time intervals in humans [Merchant2013]. The **cerebellum** is important for some forms of motor timing, such as eyeblink conditioning [Mauk2004]. It's worth noting that, despite its name, the temporal lobe is not directly related to the ability to measure time, which has led to erroneous assumptions about its role in timing.

The discrimination of time intervals improves with practice, but this improvement is specific to the trained interval and does not generalize to other scales, which supports the existence of specialized timers rather than a universal "master stopwatch" [Merchant2013]. Professionals like musicians, who require precise timing, demonstrate superior performance in various temporal tasks, suggesting an adaptation of neural circuits to the specific demands of their activity. The ability to keep a beat in music, a skill that is surprisingly rare in most animals (except for a few vocally learning birds like cockatoos), also underscores the sophistication of the sensorimotor integration necessary for complex timing [Patel2006].

Understanding the brain in relation to time extends to profound philosophical implications. Consciousness is postulated to be a mental construction, an "illusion factory," that provides an edited and delayed narrative of reality [Dennett1991], [Koch2004]. It is generated discontinuously, often with a delay of at least a third of a second from when a stimulus is detected by the unconscious brain [Libet1983]. This raises deep questions about causality and intentionality, such as whether conscious decisions are mere sensations accompanying unconscious neural processes. As for **free will**, from a neuroscientific perspective, it is defined as the conscious feeling that accompanies decisions made by unconscious neural processes. Studies have shown that brain activity can predict simple decisions before the person is aware of them [Libet1983]. The decision is ultimately the result of brain activity, which implies that the person is responsible for their actions, even if the genesis of conscious will may be subsequent to neural activation [Wegner2002].

Finally, our interpretation of the laws of physics is filtered through the intrinsic architecture of the brain. It is debated whether the perception of the flow of time, such a robust and universal experience, is a purely mental illusion or if, on the contrary, it reflects a fundamental physical feature that the current laws of physics have yet to fully explain [Carroll2010], [Rovelli2018]. The perception of the flow of time (past-present-future) does not align with certain formulations of contemporary physics, such as relativity, where all temporal coordinates exist simultaneously. This suggests that the **"arrow of time"** might be a biological construction, emerging from the brain's architecture, and not an intrinsic property of the physical universe [Reichenbach1956].

Physics offers two main theories on the nature of time: **presentism**, which posits that only the present is real, and **eternalism**, which considers the past, present, and future to be equally real within a four-dimensional "block universe" (three spatial and one temporal) [Carroll2010]. Einstein's special relativity, by demonstrating the relativity of simultaneity (two events that are simultaneous for one observer may not be for another in relative motion), provides a compelling argument for eternalism, as it implies that all moments in time are "already" laid out [Hawking1988]. Furthermore, some solutions to the equations of general relativity suggest the possibility of time travel, which also supports the eternalist view. However, the fundamental laws of physics do not assign special meaning to "now" or to the direction of time (they are time-symmetric), which contrasts with our subjective experience. The physical "arrows of time," such as the **entropic arrow** (derived from the second law of thermodynamics and the tendency for entropy to increase) [Albert2000], the **cosmological arrow** (the expansion of the universe), and the **quantum arrow** (the collapse of the wave function), explain why macroscopic events appear irreversible [Carroll2010], [Rovelli2018]. To these is added the **psychological arrow**, which manifests in our inability to remember the future.

The challenge lies in reconciling the robust subjective experience of the flow of time with the eternalist view of physics. Some suggest that this perception is a mental "illusion," a narrative the brain constructs from "snapshots" of the block universe, where memories of past moments are integrated into the "now" [Buonomano2017]. However, this explanation is complex and raises the question of whether consciousness itself can exist in instantaneous "slices" of time or if it requires a "temporal thickness" to manifest, being more analogous to music than to a static image [Rovelli2018]. Evolution, by favoring adaptations that correlate with external reality (like the perception of color or pain), suggests that an illusion as universal as the flow of time should have an adaptive value, which is difficult to explain if the flow is completely unreal on the physical plane.

If we accept the premise that cells are time-generators, it follows that each organism constructs its own temporality. This perspective has significant implications for fields like medicine (chronotherapy), artificial intelligence (through neural networks endowed with internal timers), agriculture (in relation to plant circadian rhythms), and ecology (concerning the activity rhythms of animals and microorganisms). Future research could extend beyond the brain, exploring whether connective tissues or microbiomes have their own forms of timing, or what kind of "time" artificial networks construct when provided with memory and feedback [Buonomano2017]. The fundamental question that emerges is whether time itself could be an emergent property of self-organizing systems.

In summary, the brain reveals itself as an extraordinarily complex organ that has evolved to anticipate the future, actively measuring and creating our experience of time through diverse and specialized neural mechanisms on different scales. Elucidating how the brain processes time is essential for unraveling the complexity of the human mind and, possibly, for redefining our understanding of the nature of time in the universe. Instead of conceiving of time as a pre-existing canvas onto which life is projected, an ontological re-evaluation is proposed that posits it as an invention inherent to life itself [Buonomano2017]. Time, under this view, is not a magnitude that passes independently, but a property that emerges intrinsically from the living.

A Lateral Thought

From a photon's perspective, time does not exist. Moving at the maximum possible speed, it has no perception of moving from one point to another. For the photon, space is nonexistent and the universe is frozen, with no notion of before or after. There is no journey, only an instantaneous connection between its point of emission and absorption. A photon emitted ten billion years ago and captured today by your eye has not aged an instant; for it, the journey was timeless [Rovelli2018].

This apparent paradox—that the photon does not 'perceive' motion—leads us to rethink time not as a universal entity, but as an experience that emerges tied to mass, change, and entropy. Where there is no mass and no friction, there is no becoming. Time, as we know it, is a property inherent to those of us who cannot move at the speed of light. For photons, existence is reduced to an eternal now.

THE SUPERORGANISM

In previous discussions, we've examined the adaptive heuristics that underlie the complexity of living systems, from intricate intracellular biochemical responses to decision-making in complex neural networks. Among these, social and coordination heuristics are particularly important, as they have catalyzed the emergence of one of the most sophisticated configurations in evolutionary organization: the superorganism. These systems, characterized by the functional integration of multiple individuals operating in concert, transcend the classic notion of the individual organism. They exemplify the principles of distributed cognition, projecting the functional rules inherent to a brain onto a macrobiological scale.

The concept of the superorganism was formally introduced by William Morton Wheeler in 1911 [Wheeler1911]. Based on his observation of ant colonies, he noted that they functioned as indivisible units, exhibiting a cohesion and division of labor reminiscent of a multicellular organism's structure. This seminal proposal marked a paradigm shift, suggesting that natural selection doesn't operate exclusively on individuals or genes but can also favor the emergence of adaptive units at higher hierarchical levels of biological organization. In this sense, a superorganism can be thought of as an analog to a multicellular system, where each individual organism takes on the role of a specialized cell, contributing indispensably to the maintenance and viability of the whole.

A defining feature of many superorganisms is **eusociality**, a highly evolved system of social organization predominantly seen in insects like ants, bees, and termites. This type of organization is distinguished by adults living together in groups, cooperative and alloparental care of the young, overlapping generations (which allows young individuals to interact with their parents and grandparents, facilitating the transmission of information and social learning), and, crucially, a **reproductive division of labor** [Holldobler2009]. This last feature means that only a select subset of individuals (commonly the queens) has the ability to reproduce, while others—genetically sterile or with suppressed reproductive capacity—perform essential tasks for the colony's maintenance, defense, and provisioning. This apparent reproductive altruism, which at first glance contradicts the drive to maximize individual fitness, was explained by the theory of **kin selection**, proposed by William Hamilton in the 1960s [Hamilton1964].

According to this theory, an altruistic behavior can be genetically favored if it increases the altruist's **inclusive fitness**—that is, the sum of its own direct reproductive fitness and the reproductive fitness of its relatives, weighted by the degree of genetic relatedness. **Hamilton's rule (rB > C)** states that altruism will spread if the coefficient of relatedness (r) between the altruist and the beneficiary, multiplied by the benefit (B) to the recipient, exceeds the cost (C) to the altruist. This theoretical framework has been fundamental to understanding how extreme cooperation and sterility can evolve in highly integrated societies.

The direct consequence of this evolutionary logic is the emergence of a system of highly specialized **castes**, where individuals adopt roles analogous to cell types in multicellular organisms [Holldobler1990]. This morphological and behavioral specialization allows for unprecedented efficiency in resource allocation and task execution. In **honeypot ants** (*Camponotus inflatus*), for example, certain workers transform into "repletes": individuals with a hypertrophied abdomen that function as living reservoirs for nectar and honeydew. These immobile workers hang from the ceilings of underground chambers, storing food that will be used collectively by the colony during times of scarcity [Holldobler2009], serving a function similar to that of fat cells (adipocytes) in higher organisms. The composition of the stored honey is regulated by the colony, optimizing its nutritional value and preservation.

Another classic example of functional specialization and co-evolution is that of the **Attini ants**, which have developed a form of symbiotic agriculture with a specific fungus for approximately 60 million years [Holldobler2009]. The leaves and other plant materials collected by the workers are not consumed directly but serve as a substrate to cultivate the fungus in complex underground gardens. This fungus, in turn, is the colony's sole food source. This relationship has reached such a high level of interdependence that the fungus has lost the ability to reproduce without the constant intervention and care of the ants, exemplifying a closed loop of functional co-evolution. The sophistication of this fungal agriculture includes weed control (of invasive fungi), fertilization of the substrate with chewed plant material, and prevention of fungal diseases through the secretion of natural antibiotics produced by symbiotic bacteria living on the ants.

Hyperspecialization also reaches its peak expression in **termite queens**, whose abdomens transform into a biological egg-laying machine that can produce up to 30,000 eggs a day. Entirely dependent on workers for feeding, grooming, and the transport of her eggs, the queen represents the central reproductive apparatus of a superorganism that perpetuates itself beyond the lifespan of its constituent individuals. The exceptional longevity of these queens, which can be a hundred times that of their workers, is attributed to the protection the colony provides, drastically reducing their external mortality.

To understand the intricate complexity and adaptive capacity of these systems, the **"Inductive Machine"** model is particularly useful [Holland1986]. This model describes systems that can learn and dynamically reorganize when faced with internal contradictions or environmental challenges. They use mechanisms of trial and error, or collective-level variation and selection, to stabilize functional patterns that prove successful. In insect colonies, interactions mediated by pheromones (chemical signals), dances (behavioral signals), or tactile stimuli act as channels for exploration and feedback, allowing the collective system to adjust its strategies and adopt successful behaviors that crystallize into stable behavioral rules.

Adding to this view is the concept of **"constraint closure,"** developed by Montévil, Mossio, and Kauffman [Montevil2014]. This perspective holds that living systems not only operate under pre-existing constraints (like the laws of physics or resource availability) but also generate and maintain their own functional constraints, which gives them autonomy and the capacity for self-organization. A superorganism, in this light, can be understood as an entity that organizes and maintains thermodynamic work cycles—such as food collection, territorial defense, reproduction, or social immunity—that close in on themselves, perpetuating the viability and identity of the collective system.

Social immunity in bee colonies is an illustrative case of how these functional constraints manifest at a collective level [Cremer2007], [Seeley1982]. Behaviors such as applying propolis (a resin with antimicrobial and antifungal properties) to the hive walls, mutual grooming among individuals to remove external parasites, isolating sick individuals to contain the spread of pathogens, necrophoresis (removing corpses), and, in extreme cases, the altruistic suicide of infected workers who leave the colony to prevent disease transmission, all work in concert as a **distributed immune system**. This collective system reinforces the sanitary integrity of the whole, even compensating for a weaker genetic immune endowment at the individual level. The spatial organization of the hive, as demonstrated by Baracchi and Cini [Baracchi2014], also contributes to this immunity by segregating the bees most exposed to pathogens from those who are more vulnerable or reproductively valuable.

Even without a symbolic and explicit **Theory of Mind (ToM)** analogous to that of humans, superorganisms exhibit a *functional* ToM. The waggle dance of bees, described by Karl von Frisch [vonFrisch1967], is a sophisticated mechanism for a shared representation of environmental reality: it precisely communicates the direction, distance, and quality of a food source to other bees that have not had direct access to that information. This type of representation, which allows one individual to influence the behavior of others based on acquired knowledge, can be understood as a primitive but effective form of inter-individual modeling, in line with a spectral conception of ToM that does not require the explicit attribution of mental states.

Beyond behavior, the superorganism raises profound ontological questions about biological identity. Just as personal identity in complex organisms does not reside in individual cells—which are constantly being replaced—but in the persistent dynamics of their neural networks and activity patterns, the identity of the superorganism does not lie in the mere aggregation of its individuals. Instead, it resides in its **persistent functional architecture**: the structure of the nest, the established pheromone trails, the cultivated fungus gardens, and the collective behavioral norms that guide the group's actions [Holldobler2009]. This collective identity is a product of self-organization and co-evolution.

Stuart Kauffman has argued that living systems, unlike designed machines, cannot be fully understood under the classical-physical paradigm because of their intrinsic ability to generate new functions and properties that are irreducible to prior laws [Kauffman2019]. In a **non-ergodic universe**, where possible configurations do not repeat and the exploration of the "adjacent possible" is constant, evolution is not just adaptation to pre-existing conditions but a continuous invention of new forms of life and organization. From this perspective, superorganisms are not merely the inevitable consequences of natural selection but emergent solutions that stabilize functions and capabilities that did not previously exist at the individual level, creating new units of complexity and adaptation.

Taken together, the superorganism represents an extreme crystallization of the rules and principles that underlie brain function: functional distribution of tasks, parallel processing of information, feedback-based learning through dynamic interactions, and the stabilization of behavioral patterns that optimize the system's survival. Humanity, with its intricate cultural, technological, and linguistic networks, its capacity for large-scale cooperation, and its specialization of knowledge, could be considered an **incipient superorganism**. In the face of global challenges of unprecedented magnitude—such as pandemics, ecological collapse, and climate crises—the survival and flourishing of our species may critically depend on our ability to "close our own collective constraints" and act coherently as a planetary distributed intelligence, transcending individual interests for the sake of the global system's viability.

YOUR BRAIN, THE EDITOR OF YOUR REALITY

You do not experience the world in real time. Instead, your brain continuously reconstructs reality, adjusting perceptions and memories to maintain a coherent narrative of existence [Clark2016]. But how does this redefine our notion of 'objective reality' and the very essence of human experience?

This capacity for "cerebral editing" is not a limitation but a crucial adaptive advantage. In a complex and information-saturated environment, processing every stimulus in real time and at maximum resolution would be energetically inefficient and cognitively overwhelming. The brain has evolved to prioritize efficiency and survival: by interpolating data, filling in gaps, and predicting the immediate future, it allows us to make quick decisions and optimize resources. This is a fundamental principle that resonates with the logic of the "Inductive Machine," which seeks efficiency in resolving contradictions [Putnam1966].

This editing ability is rooted in synaptic plasticity, the fundamental mechanism by which connections between neurons strengthen or weaken in response to activity. As explored in the 'Inductive Machine' model [Holland1986], the brain doesn't just reinforce successful patterns (an echo of the Hebbian principle, "neurons that fire together, wire together" [Hebb1949]). Crucially, it uses the resolution of internal contradictions as an engine to generate new 'rules' and reorganize its circuits. Thus, the reality we perceive is not a mere passive representation of stimuli, but an active and

101

dynamic construction, continuously 'edited' and refined by these synaptic reconfigurations driven by experience and the need for coherence.

Vision is our primary sensory channel, yet it is riddled with inherent limitations. Only a tiny fraction of our visual field is processed in high resolution, while the rest is actively interpolated by the brain [Dennett1991]. This aligns with the principles of Gestalt psychology, which suggest that our brain organizes visual elements into unified wholes based on proximity, similarity, continuity, and closure [Kandel2021]. In this sense, the brain's ability to fill in visual gaps and organize fragmented sensory information directly reflects these organizational laws.

Throughout the day, our eyes make three to four saccadic movements per second, during which visual perception is temporarily shut down to avoid creating disjointed images [Dennett1991]. In total, this adds up to approximately two hours a day spent in a state of temporary, and completely unnoticed, blindness.

Without the neural corrections made by the nervous system, our visual experience would be chaotic, similar to an interrupted data stream: *brrrr... brrrrrrrrr*. Instead, the brain synthesizes a stable, continuous image by fusing fragments of sensory data [Eagleman2008]. This cognitive editing capability is not limited to the visual realm; it extends deep into our perception of time, where the brain orchestrates a similar synchronization to build a coherent narrative.

Consider the act of catching a ball: the light reflected from the object reaches your eyes in nanoseconds, the sound of the impact travels through the air in milliseconds, and the tactile sensation in your hands takes about 50 milliseconds to be processed [Buonomano2017]. Each stimulus registers at a different moment; yet, instead of perceiving them as asynchronous, the brain syncs them into a unified temporal experience. This process is analogous to how TCP/IP protocols work in telecommunications, where data is broken into packets that may take different routes before being reassembled at their destination [Eagleman2008]. Similarly, the brain integrates disparate sensory signals to create a coherent narrative of the present.

However, this integration is only part of the process: the brain doesn't just reconstruct the past, it also predicts the immediate future [Schacter2007]. Imagine driving at high speed when the car in front of you suddenly brakes. If your brain processed this information in real time, your reaction would be dangerously delayed [Libet1983].

To overcome this, the brain dynamically predicts events. Using prior sensory data, it models possible trajectories and contingencies, constructing a probable future before it happens [Friston2010]. What you perceive while driving is not the immediate present, but a projected version of reality, designed to optimize split-second decision-making.

Before you consciously perceive the danger, your brain has already considered several possible responses:

- Immediate braking
- A sharp swerve
- Changing lanes

In that brief instant, multiple futures coexist in your nervous system, but only one materializes into your final action [Kahneman1979].

When you walk, you feel as though you control every step. In reality, you operate within an expanded temporal frame [Wittmann2013]:

- You process past sensory information.
- You assess the current state of your body.
- You predict the motor dynamics of your next few steps.

Before your foot even touches the ground, your brain has already issued commands for the next step and adjusted the muscular coordination for the two steps after that [Miller1960]. If an unexpected obstacle appears, like a slippery surface, your brain already has corrective mechanisms in motion. Meanwhile, the spinal cord and brainstem respond faster than your conscious perception [Kandel2021]. For example, when your inner ear detects a change in balance, it sends signals to subcortical motor centers, triggering postural adjustments in about 200 milliseconds. By the time you become aware that you've slipped, your body has already begun trying to stabilize itself.

Physiological sensations, emotions, and behavioral responses are not mere reactions to external stimuli; they are the brain's predictions about future demands [Hohwy2013]. If your circadian rhythm adapts to fixed times for eating or sleeping, your nervous system anticipates these events by releasing hormones at the appropriate moments. Emotions work on a similar principle: if you've experienced anxiety in a particular situation before, your brain may preemptively adjust your autonomic responses, preparing you for the feeling of anxiety before it even occurs.

This predictive model does not imply a lack of conscious control. While everyday decision-making relies heavily on automaticity, consciousness acts as a metacontrol mechanism [Fleming2012]. It allows us to override, adjust, and guide these automated predictions, especially in novel or complex situations.

Even though each brain builds its own subjective reality, social interaction allows us to synchronize these individual "edits." Through language, non-verbal communication, and the ability to infer the mental states of others (Theory of Mind) [Premack1978], humans are able to construct a shared, consensual reality, which is fundamental for cooperation and social life. This ability to negotiate and unify individual realities is another manifestation of our brain's adaptive intelligence on a macro level.

In essence, your perception of the world is an inferential construct, a masterpiece of cerebral editing so convincing that you experience it as unquestionable reality. What implications does this truth have for the nature of what we call 'reality' and our place within it?

THE HARD PROBLEM

Conscientia

In the contemporary fields of philosophy of mind and neuroscience, consciousness has been one of the most challenging and elusive questions. In the 1990s, philosopher David Chalmers introduced a fundamental distinction that has permeated both philosophical and scientific debates: the "easy problems" and the "hard problem" of consciousness (from the Latin *conscientia*, "shared knowledge"). The former refer to those aspects of consciousness that, although complex, are accessible to standard research methods, such as cognitive, perceptual, and behavioral processes. These can be broken down into neurobiological and functional mechanisms, such as the processing of sensory information in the brain or the formation of memories [Kandel2021]. For example, cognitive neuroscience can explain how the brain processes visual or auditory signals, or how it organizes the information that forms our memory.

And yet, for all that neuroscience can explain, a deeper mystery remains. The hard problem, according to Chalmers, lies in explaining how the physical and neural processes of the brain generate subjective experience, or "qualia" [Koch2004]. These qualitative experiences, such as the perception of the color red, the sensation of pain, or the emotion of listening to a piece of music, are not reducible to the mere activation of neural networks. This qualitative and personal aspect of experience challenges purely reductionist explanations. If, as we explored earlier, our perception of the world is an inferential construct and a "masterpiece of brain editing" [Clark2016], then the "hard problem" deepens: how does this orchestration of data and predictions give rise to the felt quality of experience, to the qualia that tinge our subjective reality with color, sound, and emotion?

The concept of this "hard problem" was formally articulated by Chalmers in his seminal 1995 paper, *Facing Up to the Problem of Consciousness*, and has since become a central axis for contemporary debate on consciousness. Over the decades, the term has gained notoriety in both philosophy and neuroscience, highlighting a crucial difference between the processes of the mind that are explainable by traditional scientific tools and those that continue to elude our understanding. Specifically, the fundamental question posed by the hard problem is how the physical states of the brain, which can be studied objectively through neuroscientific methods, can give rise to the subjective experience that characterizes our consciousness [Dennett1991].

The traditionally posed difficulty lies in the fact that while the easy problems can be addressed with established scientific methods, the question of conscious experience did not seem reducible to purely functional explanations. This explanatory gap was clearly seen when considering how the brain is capable of performing complex cognitive tasks without necessarily involving a subjective experience of them. However, recent advances in the understanding of complex systems and emergence suggest that this apparent irreducibility might be overcome [Strogatz2003]. Despite advances in neuroscience, felt experience remains a phenomenon that cannot be fully understood through traditional investigation of brain mechanisms.

To address this profound enigma, several innovative theories have emerged. One of the most influential is the Integrated Information Theory (IIT), proposed by Giulio Tononi. According to IIT, consciousness is not simply the result of a system's activity but depends on a high degree of information integration within that system [Deco2011]. In other words, consciousness appears when a system has the capacity to integrate information in a unified and differentiated manner. This theory provides a quantifiable approach to measuring consciousness, which is a significant advance in a field where objective measurement has been difficult to achieve. Imagine a puzzle: consciousness, according to IIT, would be greater the more interconnected the pieces are and the more information is integrated to form the complete picture. However, IIT has been the subject of criticism, especially regarding the lack of empirical evidence to support its postulates, as well as the difficulties in testing it in complex biological systems like the human brain.

Moving from the level of information to the very fabric of physics, another recent theory that has generated considerable discussion is the quantum consciousness hypothesis. This hypothesis, championed by Roger Penrose and Stuart Hameroff, suggests that quantum processes occurring in the microtubules of neurons may be fundamental to conscious experience. The proposal holds that the collapse of the wave function in these microtubules could be related to the emergence of consciousness, suggesting that quantum effects may play a crucial role in generating subjective experience. However, although emerging studies suggest that certain biological processes, such as photosynthesis and avian navigation, might depend on quantum effects at non-extreme temperatures, the hypothesis that the human brain leverages these phenomena—to generate consciousness—remains highly speculative [Kauffman2019]. This hypothesis faces the difficulty of conclusively demonstrating such quantum processing under conventional physiological conditions, which raises significant questions about its applicability and relevance in the realm of human consciousness.

In addition to the theories discussed, it is important to recognize the wide range of approaches that scientists and philosophers continue to explore. For example, emergent materialism and functionalism continue to offer perspectives that view consciousness as a product of the brain's organization and processes. Emergent materialism posits that consciousness emerges from the complexity of interactions among the physical components of the brain, without being reducible to these individual components [Strogatz2003]. Functionalism, on the other hand, focuses on the functional role of mental states, arguing that consciousness does not depend on the specific physical substrate (like the brain), but on the causal relationships between different mental states [Dennett1991]. Approaches like panpsychism suggest that consciousness is a fundamental phenomenon in the universe, present in varying degrees in all matter. Analytic Idealism, proposed by Bernardo Kastrup, maintains that reality is, ultimately, a manifestation of consciousness and that no material world exists independent of it.

Emergent Consciousness

Consciousness, a universal human experience, remains one of the deepest, though not inscrutable, scientific enigmas. The solution to the "hard problem" from a materialist and functionalist perspective lies in the understanding of complex systems and emergence [Strogatz2003].

Consider matter itself: subatomic particles, which at first glance might seem chaotic and without discernible properties like solidity or liquidity, organize themselves into atoms, molecules, and macroscopic structures. From this organization arise completely new properties, unrecognizable from their original components. A quartz crystal is not just a sum of silicon and oxygen atoms; its structure, hardness, and transparency are emergent properties of its collective arrangement. Similarly, a living organism is not simply a collection of cells; life itself is an emergent property of the intricate organization and dynamic interactions of those cells [Kauffman2019].

Nowhere is this principle more evident than in the brain. The human brain, and indeed every organism, stands as irrefutable evidence that the actions we—as humans—consider or label as 'intelligent' are formed precisely in that universal inductive process. We are the living testimony of how the union of fundamental blocks, the convergence of innumerable inductive agents, and the collective reaching a consensus, gives rise to intelligence. This superorganism of unprecedented complexity is not a single entity processing information, but a vast collective of interconnected processing units operating at multiple levels [Holldobler2009]. From the individual sensory receptors in the skin that detect a stimulus, to the neural networks in the brainstem that regulate breathing, each component is an integral part of a system that seeks to resolve contradictions and adapt. In this framework, specialized agents such as those responsible for language, image processing, or pattern consolidation (which in our experience manifests as rumination or memory fixation, and in simpler organisms could operate as a form of consolidating adaptive responses), contribute to this emergent intelligence we call consciousness. It is through the interaction and consensus of these agents that the global system achieves behaviors and experiences that transcend the sum of its parts.

We observe this principle of emergence in nature in surprising ways. Flocks of starlings, for example, create gigantic, fluid aerial formations that move as a single organism, without a central leader. Each bird follows simple rules of interaction with its nearest neighbors, and from these local interactions emerges a global choreography of astonishing beauty and efficiency [Strogatz2003]. Similarly, ant colonies exhibit a collective intelligence that allows them to find the shortest path to a food source or adjust the division of labor without any individual ant having an overview or an awareness of the collective goal [Holldobler1990]. These are palpable examples of how the collective is truly greater than the sum of its parts.

Even at more fundamental levels, the ability to "store data" and use that information to resolve contradictions is evident. The slime mold *Physarum polycephalum*, a unicellular organism, solves mazes and optimizes nutrient transport networks. It achieves this by leaving a slime trail that acts as an externalized "spatial memory," indicating where it has already explored. Furthermore, its internal oscillations and cytoplasmic flow adjust to reinforce the most efficient routes, an analogue of processing and "consensus" that allows the organism as a whole to "decide" the best strategy and resolve contradictions [Huang2024]. This example, on a seemingly simple biological scale, underscores how intelligence emerges from the ability of units to interact, store, and use information, even without a brain.

Most of this collective activity occurs unconsciously. Our internal systems constantly self-regulate: the heart beats, the lungs breathe, digestion proceeds, all without our direct conscious intervention. These are examples of how the collective of processing units in the body resolves complex physiological contradictions, maintaining homeostasis and enabling survival [Cannon1932]. These associations and communications occur at levels that, for the most part, do not reach the thresholds of linguistic agents or deliberate cognition, which is why we say they "remain in the unconscious." However, it is crucial to understand that the ability to process what has been experienced, to "ruminate" on past experiences or anticipate future scenarios, does not necessarily require a language module or explicit symbolic thought; it simply happens. These processes are fundamental to our existence.

This is precisely where the solution lies. From this perspective, consciousness emerges as a higher-order property of this cerebral superorganism. It is not a function of a single neuron or a specific area of the brain, but the result of massive integration and global coherence that arises from the dynamic interaction of billions of processing units [Singer1999]. The qualia, those felt qualities of experience, would be nothing more than the qualitative characteristics that emerge when the vast collective of the brain achieves a unified and highly integrated solution to the innumerable "contradictions" it continuously processes. It is the "feeling" of that coherence, the experience of a unified and coherent model of reality, both internal and external.

In essence, the "hard problem" transforms from an unfathomable mystery into a comprehensible challenge within the framework of emergent complexity. Consciousness is not a "ghost in the machine," but the most sublime manifestation that the collective is truly greater than the sum of its parts, a new characteristic that arises from the large-scale organization and interaction of living matter [Dennett1991]. This perspective allows us to approach consciousness not as an exception, but as the pinnacle of the adaptive capacity of complex systems, an emergent phenomenon that allows us not only to process reality but also to feel it in all its subjective richness.

PART III

COLLECTIVE MINDS AND HEURISTICS

WHY DO WE HAVE A BRAIN?

The question, "Why do we have a brain?" is much deeper than it first appears, inviting us into a profound reflection on the nature of life, intelligence, and organization. Could it be that the existence of our complex thinking organ boils down to a matter of organizational convenience?

It's an observable fact that countless life forms thrive without a brain structure analogous to that of animals. Plants, for instance, lack a centralized nervous system, yet they demonstrate a surprising ability to communicate and respond to their environment, challenging the notion that intelligence is exclusive to neurons [Mancuso2015]. Octopuses, with their distributed nervous system, are another prime example of how intelligence can emerge in unconventional ways [GodfreySmith2016]. This leads us to question whether the brain is truly the exclusive domain of intelligence, or if intelligence arises, more fundamentally, from cellular collectivism itself. In fact, organisms like slime mold show that cognition, at its root, is a manifestation of life's decentralized organization.

This brings us to a fundamental claim: intelligence, and by extension consciousness, is an emergent property of cellular collectivism [Kauffman2019]. It's not an attribute exclusive to neurons but a phenomenon that arises from the organized interaction of any community of cells, whether they are neural or not.

In a world of constant change, survival depends on the ability to anticipate and effectively respond to environmental cues. Plants, algae, and photosynthetic bacteria, though immobile or structurally simple, have colonized nearly every ecosystem on Earth. Many

114

depend on external agents to spread their genes, and attracting these vectors requires sophisticated strategies: colors, fragrances, rewards. Some orchids take it a step further, disguising their shape, color, and scent to mimic female insects. They trick the male insect into mating with the flower, which then transports the pollen without receiving anything in return—a highly efficient, minimum viable product in evolutionary terms.

This type of deception, a remarkably precise evolutionary heuristic, triggers key responses without requiring neural processing. We could almost describe it as a rudimentary mental model: an evolutionary prediction about another organism's behavior. It's a kind of botanical Theory of Mind—lacking intent or consciousness, but full of adaptive efficiency. Nature, as Maupertuis postulated in the 18th century and Feynman later formalized, favors the principle of least action: the most energy-efficient solution [Feynman1965].

The *Mimosa pudica* folds its leaves upon contact. Some plants emit volatile compounds to warn their neighbors or attract predators to fend off their attackers [Kessler2001]. Bacteria activate collective mechanisms like quorum sensing, acting only when they reach a critical mass [Bassler2002]. These examples show that intelligence, or at least adaptive efficiency, predates the brain. Cellular organization is enough to produce highly functional and efficient responses to changing stimuli.

In this context, the brain is not a command center, but an organizational convenience. It's a sophisticated predictive network in service of the body. Its primary function isn't "to think," but to coordinate, regulate, and adjust. In neuroscience today, the brain is understood as a prediction machine [Clark2016, Hohwy2013]. Locked inside the skull, it only accesses the world indirectly. What we perceive is an active inference: a blend of prior expectations and incoming sensory signals. This Bayesian logic, which echoes the principles of an inductive machine seeking to minimize error, is not unique to humans [Friston2010]. For all organisms, to act is to predict and minimize error in the face of the unexpected.

Consider animal defense strategies. An elk, locking antlers with a rival, adjusts its stance based on the opponent's every move. A bison decides whether to flee or charge based on its "best guess" about the threat. The rhinoceros beetle tosses its rivals with surgical precision. The mantis shrimp predicts its prey's position before launching a strike so fast it creates cavitation bubbles. Each one acts to reduce its own uncertainty.

Chemical arsenals follow the same logic. A rattlesnake's warning predicts its opponent will back off; if that prediction fails, it strikes. The stonefish first relies on its invisibility. If that fails, its venomous spines come into play. The blue-ringed octopus and the cane toad also operate under this mechanic of anticipation and response.

Camouflage and evasion refine this strategy further. A zebra's stripes confuse a lion's perception. A chameleon adjusts its coloration to hide or to hunt. An octopus releases a cloud of ink as a last resort when its camouflage fails. All of these are attempts to minimize the other's prediction accuracy, or to maximize their own.

The brain is not the source of intelligence but a sophisticated extension of more ancient strategies. Like an extension of cellular metabolism, it acts to maintain homeostasis and avoid surprises that threaten the organism's integrity [Cannon1932, Friston2010]. Through action, we seek to confirm our predictions. We also regulate our internal state; interoception is a form of active inference [Seth2021]. We are a conscious self because we are a system of self-fulfilling prophecies.

Within this framework, these internal representations of the environment and of ourselves—which we know as mental models—are built from past experience and refined by active inference [Johnson-Laird1983]. They allow us to anticipate consequences, simulate alternatives, and choose courses of action without having to physically execute them. They aren't exact maps but functional, adaptive approximations that guide our decisions and help us minimize error in uncertain scenarios. From the zebra predicting its predator's perception to the human planning a difficult conversation, we all act guided by mental models that shape our interactions with the world. Understanding these models allows us to appreciate that cognition, like life itself, is not centralized or rigid, but distributed, flexible, and deeply rooted in the evolutionary history of living matter. It is precisely the analysis of these predictive architectures, these mental models, that will be the common thread of the next part of our exploration.

IN THE BLINK OF AN EYE

I've always believed that sometimes we fail to see things not because they are invisible or nonexistent, but simply because we haven't learned how to see them yet. This idea reminds me not only of dark matter but of the entire universe of particles. Here, however, we'll talk about a different kind of blindness: the one that stems from the rhythm at which we perceive the world.

In the *Star Trek: The Original Series* episode "Wink of an Eye," the crew of the *Enterprise* responds to a mysterious distress call from the planet Scalos. What they find is a desolate city, seemingly empty and lifeless. Soon, however, a series of inexplicable events begins to plague the crew: sudden system failures, objects moving on their own, and strange buzzing sounds that register as if something—or someone—is moving at a dizzying speed beyond their comprehension. Most unsettling are the sudden disappearances of crew members, who seem to vanish into thin air. It's only when Captain Kirk, by a twist of fate, is hyper-accelerated into their timeline by drinking the planet's water that a chilling truth is revealed. From this new, vertiginous perspective, he discovers the Scalosians: beings whose entire existence unfolds at such an extreme velocity that humans, at their normal pace, appear to be frozen solid. For the Scalosians, this speed is not an anomaly but their everyday reality, and humans are mere static figures, oblivious to the vibrant life buzzing all around them. Their queen, Deela, with calculated coolness, reveals her plan to use the *Enterprise* crew as a genetic stock to preserve her sterile civilization, all while the humans remain completely unaware of their presence, unable to even perceive them.

This narrative vividly illustrates the coexistence of hidden realities, imperceptible from a different timescale. In a similar way,

118

humanity has traditionally viewed the plant kingdom as a passive, silent backdrop for the drama of animal life—a perception shaped by our own temporal scale. At the speed we move, plants seem motionless; they barely perceive us as fluctuations in electric fields, and we rarely notice their movements. In the frame of reference of a plant, we are the buzz, the blur that strikes with force, the fast-movers.

Today, researchers are transforming our preconceived notions, their observations revealing a world of complexity, sensitivity, and communication [Mancuso2015, Trewavas2005]. Plants, despite lacking a heart to pump blood or a centralized brain to process thoughts, have evolved extraordinarily original solutions to perceive, interact with, and survive in their environment. These discoveries not only compel us to admire their resilience but also invite us to learn from their sophisticated intelligence, blurring the boundaries we once believed were firm between the human, animal, and plant worlds [deWaal2016].

What distinguishes plants from most animals is their sessile nature. While mobile organisms can evade threats or actively seek resources, plants are intrinsically anchored to their substrate. This inherent immobility, far from being a limitation, has driven the evolution of extraordinarily sharp sensory capabilities [Trewavas2005]. The need to anticipate, perceive, and react to their environment with precision is critical for the survival of an organism that cannot run from danger.

This heightened sensitivity doesn't rely on a centralized brain. The brain, in essence, is a network of specialized cells (neurons) that transmit electrical signals. Plants demonstrate that information processing can be achieved through alternative biological architectures [Levin2021]. They use a vascular system, functionally analogous to an animal's circulatory system, and connections that facilitate the transmission of electrical signals and hydraulic pressure waves for internal communication. For example, an insect bite on a leaf can induce an electrical signal that propagates through the stem, albeit at a considerably slower speed than an animal nerve impulse (centimeters per minute versus meters per second). This speed, while slow by our standards, is congruent with the metabolic rhythm and lifestyle of a sessile organism. Simultaneously, a hydraulic pressure wave can reach the roots, alerting the entire plant to the threat and enabling coordinated responses, such as closing stomata to conserve water or reallocating energy toward defense mechanisms.

Contemporary research reveals that plants perceive their environment through senses that, while operating through different mechanisms, exhibit surprising functional analogies to animal senses.

Genetically, plants possess homologs of animal genes involved in hearing, particularly those that code for the vibratory cilia of the inner ear. In plants, these genes contribute to the formation of root hairs, crucial for water absorption. Nevertheless, the ability to perceive sound vibrations is becoming increasingly evident. Roots, for example, exhibit positive hydrotropism toward low-frequency sound sources (100-1000 Hz), which could indicate the presence of water, and they grow away from higher frequencies [Mancuso2015].

A remarkable finding is the work of Lilach Hadany, who demonstrated that flowers can "hear." The flowers of the evening primrose (*Oenothera drummondii*), upon detecting the specific buzz of a bee, increase the sugar concentration in their nectar within minutes. The flower's concave shape acts as a parabolic resonator, vibrating in response to the frequencies of its pollinators while ignoring other sounds. This mechanism represents remarkable energy efficiency, as the reward is offered only when the pollinator is present. In addition to perceiving sounds, plants also "emit" acoustic signals. Under stress conditions, such as drought, plants like tomatoes emit ultrasonic clicks, inaudible to the human ear [Mancuso2019]. These sounds carry information about the plant's physiological state, suggesting a language of stress that could be exploited in agriculture for the early detection of crop problems.

Each plant species emits a unique olfactory signature, composed of a complex mix of volatile organic compounds (VOCs). This capability is not merely passive. The dodder plant (*Cuscuta*), a parasite devoid of roots and photosynthetic ability, exemplifies this active chemo-perception. To survive, it must locate a host. High-speed observations reveal how the dodder "sniffs" its environment, extending its stem and selecting its victim. It is capable of discriminating between different plant species and, even more surprisingly, can distinguish between a healthy and a sick host, rejecting the latter. The dodder perceives the chemical profile of the plant that will offer it the best sustenance. Plants can also detect their predators through VOCs. For example, the tall goldenrod (*Solidago altissima*) can perceive pheromones emitted by the parasitic gall fly (*Eurosta solidaginis*) and, in response, activate its defenses, such as postponing its flowering to avoid an attack on its seeds [Kessler2001].

The perception of light in plants is ubiquitous and multifaceted. Unlike human eyes, which capture a limited range of the visible spectrum, every plant cell is a photoreceptor. Plants "see" not only the visible spectrum but also wavelengths invisible to us, such as ultraviolet and far-red light [Cashmore1999]. This information is not translated into visual images but is integrated directly into physiological processes like germination, growth, and flowering. Beyond photosynthesis, plants use light for communication. They reflect light in the near-infrared spectrum, a wavelength they use to "perceive" their neighbors. A plant can determine if it is surrounded by competitors and adjust its growth pattern accordingly, for instance, by reducing branching to optimize resource allocation in a competition for light.

The ability of plants to process information extends to concepts traditionally associated with higher cognition, such as memory and spatial awareness. Trees exhibit a form of environmental memory; research has shown that a tree exposed to unusual wind gusts "remembers" the experience [Gagliano2014]. In response, it produces more wood to reinforce its trunk, preparing for future storms. The curve at the base of a leaning tree is a physical manifestation of the memory of a past event. Furthermore, plants possess a sense analogous to proprioception: the ability to perceive the position of their own body in space. Even in an environment where light is diffuse and the effects of gravity are nullified by rotation, a tilted plant is able to right itself. This is attributed to specialized cells containing statoliths (starch grains), which, like tiny sediments, settle due to gravity, informing the plant of its orientation and allowing it to correct it.

The intelligence of plants is manifested in their decentralized structure. The root apexes—the very tips—are centers of intense electrical activity and exhibit a disproportionate consumption of oxygen, similar to neuronal activity in an animal brain [Mancuso2015]. These thousands of apexes explore the subsoil, communicating and coordinating for the acquisition of water and nutrients. They function as a collective intelligence, a swarm system without a centralized leader, where the group operates as a coherent entity.

This network is further complexified through symbiosis with fungi, forming mycorrhizae. These fungal filaments intertwine with tree roots, expanding their reach thousands of times and creating a vast underground network, often dubbed the "Wood Wide Web" [Simard2021]. Through this network, trees not only transport water and minerals but also share carbon and, therefore, energy. It has been documented that one tree can transfer up to 40% of its carbon to a neighboring tree, even of a different species. Larger, mature trees can nurture younger individuals growing under their canopy, and a healthy tree can assist a stressed one. This system of regulated exchange and subterranean communication reveals that a forest is not merely a collection of competing individuals, but an interconnected, cooperative community.

Beyond sensory perception and memory, plants display a strategic sophistication that challenges our traditional conceptions of

122

intelligence. They have developed mechanisms that not only allow them to survive but also to radically influence their environment, orchestrating the help of others or modulating their behavior in unexpected ways.

Orchestrated Coherence

In ecosystems, some of the most sophisticated strategies come not from predators, or even from mobile organisms, but from plants. Motionless and vulnerable, unable to flee, they have developed silent yet effective tactics to orchestrate help from others. They don't fight directly; they design the conditions for others to fight for them.

When a caterpillar attacks a corn plant, it doesn't respond with thorns or poisons. Instead, it emits a chemical signal, a specific volatile compound that summons a parasitoid wasp [Kessler2001]. This small, efficient predator injects its eggs into the invading caterpillar, which ends up being eaten from the inside out. Thus, the corn plant doesn't defend itself: it coordinates. Its intelligence lies in the precision of the signal, in the exact way it activates a beneficial response from the ecosystem.

The oak tree, for its part, deals with caterpillars even more elegantly. It emits odors when it detects attacks on its buds, and these scents are interpreted by insect-eating birds like the blue tit as a promise of future food. The birds respond not only by eating the caterpillars but also by building their nests near the tree, creating continuous defense. The oak doesn't fight: it creates an environment of anticipated protection [Engelberth2004].

These strategies are not based on direct control or confrontation. They are an example of what we might call "orchestrated coherence": a way of influencing the world without needing to dominate it, using signals that align other actors for mutual benefit.

123

Antifragility

In some corner where nature unfolds its subtlest strategies, a wild tobacco plant wages a silent battle for survival and propagation. Every night, its white flowers exhale a sweet perfume, an ethereal invitation to the tobacco hawk moth. The moth, punctual in its ancestral dance, feeds on the nectar and, unintentionally, transports the pollen that ensures the plant's genetic continuity, mixing lineages and strengthening the species with the vitality of diversity. However, this act of collaboration carries a shadow: the moth lays its eggs, and from them emerge voracious larvae, caterpillars that feed on the very leaves that sustained their parent. It is a double-edged pact, where the need for diversity is pitted against the threat of annihilation.

When the insatiable appetite of the caterpillars threatens to overwhelm the plant's ability to recover, the tobacco plant deploys an unexpected strategy, a silent lesson in adaptation. Its nocturnal flowers close, and at dawn, they open again, offering their nectar to a new visitor: the hummingbird. Small, agile, and free of larval consequences, the hummingbird also pollinates, albeit with a more limited genetic reach. It doesn't bring the voracity, but neither does it bring the genetic richness of the wandering moth. The plant, facing a specific threat, modulates its behavior, sacrificing one benefit for the sake of immediate survival.

But the story doesn't end there. The presence of the hummingbird, with its liveliness and constant flight, attracts other inhabitants to the tobacco plant's ecosystem: small birds that share the hummingbird's biome. These birds discover an easy and abundant food source in the young, tender caterpillars. Thus, the plant, through an initial adaptation to avoid destruction, triggers a cascade of interactions that benefit it in an unforeseen way. The initial threat becomes bait for natural biological control. Even the caterpillars that manage to evade the birds' beaks do not escape unscathed. By feeding on the tobacco leaves, they incorporate the toxicity of nicotine into their bodies, a chemical defense that makes them less palatable to future predators, at least those of their size. Adversity, the selective pressure, even shapes the threat itself, endowing it with a stress-induced individual resistance.

This ecological drama fits the strategy of Antifragility, where adaptive systems not only react to stress but leverage it as information and a pathway for development [Taleb2012]. The revealed vulnerability gives rise to a new form of resistance. The goal is not just to withstand the impact, but to transform from it, acquiring defenses, alliances, and a clearer understanding of the environment.

Mimicry

In an environment of continuous transformation, the persistence of photosynthetic organisms lies in their ability to react appropriately to environmental stimuli. Despite their immobile nature, they have colonized every habitable corner of the planet. To ensure their continuity, reproduction is essential. Many rely on certain "providers" that move freely, such as insects, birds, and mammals—in short, pollinators of all kinds. But for a pollinator to move from one flower to another, it has to find something of interest. To attract those who spread their lineage, plants employ extraordinarily diverse resources. Some flowers display striking colors, exhale intoxicating fragrances, or excrete succulent nectars, and it works: as soon as they bloom, pollinators flock to them, laden with the fertilizing pollen they have collected from other flowers of the same species.

But some flowers go even further. The orchids of the genus *Ophrys* have developed an even more subtle strategy. Instead of investing resources in costly nectar or perfumes, they have perfected a disguise. They bear the colors, shapes, and even the scent of female bees. Their deception is so convincing that the males, confused, copulate with the flower, believing they are mating with a partner. There is no reward, no real sexual contact, but there is reproductive success: in the process, the male transports pollen to another flower of the same type. This is, without a doubt, one of nature's most elegant Minimum Viable Products (MVPs). A "minimum viable product" that, by simulating a reward without actually offering one, manages to validate its function with remarkable efficiency [Ries2011]. The orchids don't develop the entire traditional floral apparatus or invest in nectar. They present only the minimum necessary to trigger a key response: the insect's attention.

The examples of plant intelligence we've analyzed—sensory perception, environmental memory, ecosystem orchestration, antifragility, and mimicry—challenge us to move beyond an anthropocentric perspective of cognition. They force us to reconsider the notion of "intelligence" and, fundamentally, that of "mental models" [Johnson-Laird1983]. While we often conceive of these as internal, conscious representations that explain the world, plants operate with "*mental models*" that offer a radical and, in many ways, more resilient alternative.

First, their "models" are profoundly decentralized. There is no single brain concentrating information; instead, a plant's "understanding" of the world emerges from a distributed network that spans from its root tips to its vascular system and cellular chemical interactions. Each part contributes to a collective intelligence operating without a central command [Kauffman2019]. Because this intelligence *is* the whole plant, it is also inherently ecosystemic and holistic. Unlike human models that often focus on the individual in a closed system, a plant's model inherently integrates its web of interdependence with insects, birds, fungi, and the soil itself. It "knows" its survival depends not just on itself, but on its ability to orchestrate the world around it.

This intelligence operates not through domination, but through signaling and attunement. Its "knowledge" isn't translated into complex thoughts, but into the emission and reception of signals— volatile, chemical, electrical, vibrational—that connect with other actors, orchestrating behaviors for mutual benefit. This is an intelligence of subtle influence, and it is remarkably efficient. The *Ophrys* orchid is the epitome of this principle, a natural "Minimum Viable Product" [Ries2011]. Its strategic "model" doesn't invest in superfluous resources like nectar but focuses on the minimal expression necessary—a convincing disguise—to achieve its reproductive goal. It is a striking contrast to the human tendency for over-engineering, revealing a deep wisdom in simplicity and effectiveness.

Finally, this entire approach is built on antifragility and geared for long-term adaptation. Plant systems are not designed to merely resist stress, but to leverage it. Like the tobacco plant that modulates its flowering in response to a predator, or the tree that reinforces its trunk after a storm, their "models" are designed to gain from disruption, treating adversity not as a failure, but as a source of information that refines and strengthens their adaptive strategies [Taleb2012]. While their "speed" is slow on our timescale, it is this patient, resilient intelligence that has allowed them to thrive for millions of years, far surpassing the longevity of many animal species, including our own.

Today, we have new technological and conceptual tools, from real-time observation systems to the analysis of chemical signals and collective behavior patterns. This allows us to focus our attention on forms of intelligence that do not seek to imitate human structure but instead reveal alternative principles of organization, perception, and adaptation.

Non-human living beings offer us a wide range of collective behavior patterns. These allow us to analyze what we might call Cognitive Heuristics or Cognitive Models, which from an anthropocentric perspective we call "Mental Models" [Johnson-Laird1983]. Unlike models centered on a single point of control, these heuristics are not articulated from a command center but emerge from distributed networks. In these networks, information flows and is transformed into adaptive decisions.

This shift in focus is not just an epistemological expansion but a profound reconfiguration of our frames of reference. Plant intelligence, for example, challenges us to rethink what it means to perceive, communicate, or remember, and invites us to imagine an ecology of diverse minds [deWaal2016]. Far from being a mere curiosity, this perspective points toward a future in which our technologies, organizational systems, and sustainability strategies could be inspired by non-anthropocentric forms of intelligence, offering new paths for coexisting and co-evolving with the rest of the biosphere.

THROUGH THE LOOKING-GLASS

"Don't go, Dory," says Marlin.

"If you go... if you leave, I'll remember less. My memory is better with you," Dory replies, with a mix of tenderness and desperation.

Thus begins one of the most memorable scenes in *Finding Nemo*, where a fragile, forgetful blue tang fish reveals something profoundly human: the fear of losing what gives life meaning and coherence. But what if Dory wasn't just an animated fantasy? What if fish were, in fact, one of the most compelling pieces of evidence against our entrenched anthropocentric bias? In the depths of our oceans, among corals that are living cities, schools that move like a single thought, and currents that carry the secrets of the abyss, dwell beings whose inner world has remained, until now, veiled by our prejudices. This essay plunges into the depths of aquatic cognition to reveal the echoes of a universal Inductive Machine, demonstrating that the principles of intelligence reside not in the cerebral cortex, but in the fundamental logic of life itself [Kauffman2019]—a logic that traverses evolutionary time and manifests in any being that confronts uncertainty with a flexible strategy.

For generations, we have repeated the claim that fish possess an ephemeral memory—a belief as persistent as it is wrong, often wielded as a convenient justification for mistreatment and indifference. However, this myth crumbles in the face of an aquatic Inductive Machine in full operation. As cognitive biology has shown, goldfish are not only capable of remembering solutions to mazes for months, but they also build complex cognitive maps of their environment, changing their navigation strategies as they gain experience [Salas2006]. This is not a simple act of recall, but a process of active reorganization in the face of adaptive error: the system detects a contradiction—a route that no longer works—and generates a new heuristic to resolve it. Like any Inductive Machine, it reorganizes its behavior through successive approximations until it generates an effective, low-cost solution. Even under ambiguous conditions, these fish demonstrate a surprising ability to generalize experiences, anticipate rewards, and discriminate between contextual stimuli.

In another astonishing experiment, a group of goldfish learned to recognize a Bach toccata, distinguishing it from Stravinsky's *The Rite of Spring* to receive a reward [Shinozuka2013]. The ability to discern such complex acoustic patterns reveals a level of processing that defies any notion of simplicity. And on the reefs, over 980 known species communicate through sounds, vibrating their swim bladders with such precision that they can identify each other by their unique acoustic signature. A submarine symphony that we failed to hear, but that has always been there. These acoustic exchanges are not random but are structured by specific contexts: courtship, warning, navigation. They reveal an emergent, proto-functional grammar, yet to be fully deciphered, but one that points to the existence of broader symbolic coding systems.

Practical intelligence also abounds, manifesting in heuristics of adaptive efficiency. The tuskfish uses a specific rock as an anvil, returning to it repeatedly to crack open the hard shells of mollusks [Giacomo2019]. This behavior—the deliberate use of tools—is so rare in the animal kingdom that it has been observed only in an elite group of species, like primates and corvids. The fish is not simply reacting; it is applying a resource-optimization heuristic that resolves a contradiction (the hardness of the shell) with remarkable energy economy. As in any adaptive system, contradiction drives reorganization. This implies motor planning, selection of the optimal site, controlled repetition of the gesture, and spatial memory of the exact breaking point—all without the need for a prefrontal cortex.

Even more amazing is the archerfish, which hunts from the water by firing a jet that knocks insects off overhanging branches. To achieve this feat, its nervous system must execute a series of complex calculations: estimating distance, predicting the ballistic trajectory of the water droplet, and, crucially, correcting for the distortion caused by the refraction of light between water and air [Dill1980]. In essence, its nervous system acts as a "reality editor," constructing a predictive model that resolves the contradiction between the apparent and real position of the insect. The precision of this heuristic is such that, in a recent study, these fish proved capable of recognizing human faces, correctly distinguishing one person from dozens of strangers with over 80% accuracy [Newport2016]. Here, too, the Inductive Machine is at work: progressive selection of relevant patterns, reinforced by reward. This ability to adapt their shots to the size, shape, and speed of their prey suggests the existence of a dynamic statistical foundation built into their sensorimotor architecture.

And then there is the mirror. That litmus test, that portal through which we seek hints of self-awareness. A small cleaner wrasse from the Pacific (*Labroides dimidiatus*), with a brain weighing just a few grams, was exposed to its own reflection. In the experiment, a colored mark was placed on a part of its body that it could only see in the mirror. The phases of its reaction were a manual of cognition in development: first, aggression toward the "other" fish; then, unusual behaviors, like swimming upside down in front of the mirror, as if exploring the contingencies of that reflection; and finally, the revealing act. After observing the mark in its reflection, the fish swam to a nearby surface and repeatedly scraped the exact spot on its body where the mark was, a behavior that unequivocally indicates it recognized the image in the mirror as its own [Kohda2019]. This fish crossed a cognitive threshold that, until recently, we believed was reserved for great apes, dolphins, elephants, and humans. The "self-modeling" heuristic seems to have also emerged in the coral reefs. It is possible that what we call self-awareness is, at its core, a result of a system's ability to generate a representation of its own internal state projected externally.

The diplomacy of this same cleaner wrasse is, perhaps, one of the clearest manifestations of a functional Theory of Mind in a non-primate. At the "cleaning stations" on the reefs, these fish act as submarine barbers, removing parasites from their "clients." But their behavior is anything but automatic. They diligently attend to valuable clients—those who have more options to go to another cleaning station—and they serve predators with exquisite care to avoid retaliation. Sometimes, they cheat, biting off a piece of the client's nutritious mucus. When this happens and the client startles, the cleaner pursues it and offers a tactile "massage" with its fins, a gesture that appears to be an apology to retain its business. Most fascinating is the "audience effect": when other fish are watching, the cleaners are significantly more honest [Bshary2006]. This behavior is so sophisticated that it forces us to consider it a predictive model of the other. The fish doesn't "think" about its reputation in human terms, but its nervous system acts *as if* it understands the social consequences of its actions, adjusting its strategy to maximize long-term benefit. The simulation of otherness is not reflective, but it is functional. Its conduct is modulated in real-time based on an implicit social framework, revealing a form of strategic cognition that borders on the proto-ethical.

This complexity extends to the inner world of emotions. Emotion is not a human luxury. Fish possess nociceptors and brain regions analogous to the amygdala and hippocampus in mammals, structures involved in processing pain and emotional memory [Sneddon2003]. The behavioral evidence is equally powerful. The zebrafish, a social animal, will freeze upon seeing one of its own in a state of evident distress. It becomes anxious. This is a rudimentary form of empathy, an emotional contagion that strengthens group cohesion. On salmon farms, an even more disturbing phenomenon has been documented: fish subjected to chronic stress and overcrowding stop eating, isolate themselves in a corner, and float apathetically, showing all the signs of what neuroscience calls a depression-like state [Hjeltnes2017]. This is not a metaphor; it is the neurochemical response of an overwhelmed system. The affective heuristic can also collapse.

When the environmental reinforcement network disappears, organisms display energy-saving strategies that can become states of functional withdrawal.

Affective bonds, the force that weaves societies together, also have their echo underwater. In monogamous species like cichlids, the pair collaborates in defending territory and caring for their young. The forced separation of these pairs has measurable emotional consequences. In an ingenious experiment, it was shown that female cichlids, after being separated from their mates, became more "pessimistic," taking longer to investigate ambiguous stimuli that could lead to a reward [Oliveira2011]. It was as if hope had vanished with the loss. They felt the absence. The emotional heuristic not only exists; it is altered by loss. This emotional plasticity, observed in fish with relatively simple nervous systems, reminds us that a bond does not need language to exist: the regularity of interaction is enough to shape the spirit.

For centuries, we have looked at fish as silent, simple, cold beings. As living decorations or mere resources. But they are not. They have memories. They communicate. They solve problems. They use tools. They choose. They remember faces. They form bonds. They suffer. And some, through the looking-glass, even recognize themselves. In the end, the submerged mind of the fish brings us back to the central question of this work. It shows us that the Inner Architect has no preferred form. It builds with neurons, with chemistry, with pressure, and with light. Recognizing intelligence and sentience in a fish is not just an act of empathy; it is an act of epistemic humility. It is understanding that the mind is not what we *are*, but what life *does* to persist. As Dory said with her childlike wisdom, "When I look at you, I'm home." Perhaps, by seeing them with new eyes, we, too, will remember our own.

133

BLUE-BLOODED ALIENS WITH THREE HEARTS

Beneath the blue veil of the world's oceans, in a realm of pressure and gloom where time seems to dissolve into eternal currents, lives a creature that feels torn from a parallel evolutionary dream. It is a being older than the dinosaurs, whose intelligence was forged in a biological matrix entirely alien to our own, having followed a divergent path for more than five hundred million years [Kröger2011]. This is the octopus: an organism whose fragmented nervous system, extreme bodily plasticity, and sophisticated behavior compel us to rethink everything from comparative biology to the philosophy of mind [GodfreySmith2016]. Like the fish and plants we have explored, the octopus embodies a profound principle: that intelligence is not a form, but a function—a strategy for persistence in the face of contradiction. It is a mind that developed outside the linear patterns we associate with cognitive evolution, yet it achieves levels of adaptive problem-solving so refined they defy our traditional taxonomies of intellect.

The octopus belongs to the cephalopods—literally "head-foot"—a description that already hints at its liminal nature. Unlike its armored mollusk relatives, the octopus surrendered its shell and gambled on agility, behavioral adaptation, and a distributed mind. This chosen vulnerability transformed its body into a fluid, hydrostatic system without a skeleton, capable of slipping through impossible crevices. Its only hard part is a sharp beak, an ancestral testament to its lineage, which imposes the sole limit on its capacity for metamorphosis. In this renunciation of external armor in favor of flexibility, a certain logic is already implied: it is not rigid protection that guarantees survival, but the ability to vary without breaking, to respond without shattering.

But its body does not just fold; it transforms. Its skin is a high-resolution biological screen, governed directly by its nervous system. Through thousands of chromatophores per square centimeter, it can alter color, contrast, pattern, and texture in milliseconds [Mäthger2009]. This camouflage is not mere defense; it is a visual language. Octopuses communicate through pulses of color, vibrations of light, and cutaneous contortions. They send signals, encode intent, narrate tension. The skin becomes a semantic surface. What changes in the body is a reflection of a strategy: to flee, to negotiate, to attract, to simulate. In them, the heuristic of camouflage has mutated into an expressive heuristic, a kinesthetic grammar of light and form. It is a semiotics of the flesh, where color, rhythm, and texture participate in a language that is not spoken, but is manifested with the eloquence of a voice.

For a long time, they were considered solitary, almost asocial beings. But recent observations reveal forms of selective, territorial, and ritualized interaction. In certain regions, such as the reefs of Indonesia, communities of algae octopuses have been documented sharing spaces, competing, and learning from one another [Scheel2017]. They exhibit a tactile diplomacy, a recognition of hierarchies, and a choreography of gestures functionally reminiscent of the interactions of social fish at cleaning stations. Interspecies cooperation has even been observed: octopuses and coral groupers hunt in pairs, combining their complementary skills. The fish locates; the octopus retrieves [Vail2013]. It is an alliance that emerges without language, but with a shared model of the world. This is a raw form of interspecies social cognition, relying on adaptive regulation and a pragmatic modeling of the other. And when its partner fails to cooperate, the octopus has been known to punch it with a tentacle—punishment emerging as a mechanism for interspecies regulation.

In the deep waters off the coast of Costa Rica, true abyssal nurseries were discovered: hundreds of females of a new species congregate near hydrothermal vents [Giron-Nava2023]. The heat accelerates the development of their eggs, shortening a process that could otherwise take years in the cold. There they remain in a silent vigil, sharing space, tolerating proximity, and rewriting what we thought we knew about the sociability of these beings. It is a collective adaptive system, a heuristic of environmental synchronization forced by an ecological contradiction: the thermal scarcity in the darkness of the abyss. This scene of communal motherhood, without language or embrace, but with an almost ceremonial synchrony, redefines the concept of a bond.

The intelligence of the octopus is not just advanced; it is radically different. With the largest brain-to-body ratio among invertebrates, it possesses a decentralized nervous system: two-thirds of its neurons are in its arms [Hochner2006]. Each limb has its own chemical and tactile sensors and operates semi-autonomously. The octopus does not think with its head; it thinks with its body. This neural architecture allows for a simultaneity of actions, a heuristic redundancy that enables it to explore, manipulate, test, and adapt without central oversight. It is a form of embodied intelligence, as if each arm were an inductive machine testing, evaluating, and adjusting in real time. This suggests they possess an integrated, predictive body model that does not depend on a spine. Each arm is a cognitive probe, a multisensory interpretation module that rehearses the world as it interprets it.

They learn without parental instruction. An octopus is born alone, vulnerable, and what it becomes is built from direct experience. They learn to use tools, like coconuts or shells [Finn2009], to solve mazes, and to open jars with screw-on lids. This demonstrates spatial memory and an adaptive flexibility—a heuristic of progressively adjusting their patterns of action. They show evidence of associative learning and, possibly, dream-like consolidation. During active sleep, their skin reproduces the patterns of the day, as if reviewing experiences—a pigmented REM cycle. They have been observed learning by watching their peers, a form of indirect experience transfer that relies on simulating observed strategies. They also play: manipulating objects for no apparent reason, squirting jets of water, and repeating movements [Mather2015]. Play, as a cost-free rehearsal, reveals itself as a heuristic of self-generated simulation.

From the tiny wolf octopus to the Pacific giant, from the lethal blue-ringed to the ethereal Dumbo of the abyss, these beings are living manifestos of a non-vertebrate intelligence. As with fish, every action is not a reflex, but a calculation. As with plants, every gesture is a response to pressure. They are architects of their environment, designers of their camouflage, curators of their memory. They live short lives, yet learn so much. The paradox of their life cycle, with no parental teaching, suggests that intelligence can emerge even in lineages that do not rely on direct social transmission. Their body is a dress rehearsal of the possible: extreme plasticity, hierarchical multisensory perception, local and global memory, accelerated individual learning. In each tentacle, a question; in each flicker of color, a hypothesis; in each curve of their dance, an anticipation.

In many ways, they are the true aliens on Earth. Not because they are strange, but because they are convergent. They reflect back to us what intelligence could be if it were freed from our particular form [GodfreySmith2016]. And perhaps that is why, in observing them, we discover not just another mind, but another way of organizing the cognitive. A fluid, tactile, mutable way. A distributed mind that resolves contradictions through embodied exploration. If every mind is an architecture of possibility, that of the octopus is a liquid architecture: it slips through crevices, inhabits the cracks, and forgoes collective memory to intensify individual memory. At the threshold between the known and the unthought, the octopus awaits. It watches us with its horizontal pupils, responds with its curious tentacles. And we, only now, are beginning to recognize its gaze.

BRIDGES OF INTELLIGENCE

We have journeyed to the heart of identity, discovering it not as a solid essence, but as a persistent pattern: a symphony of self-organizing activity that sustains itself in the ceaseless flow of time [Kauffman2019]. But the melody of a living being never resonates in a vacuum. Every organism, from a bacterium to a primate, is immersed in an ocean of other melodies, other patterns, other identities. The fundamental question, then, is no longer just how a single pattern persists, but how it interacts with others. What happens in the space between minds?

In that space, which only appears to be empty, life creates its most intricate works. The Inductive Machine of evolution acts not only internally but externally, forging cognitive bridges [Holland1986]. It is here that intelligence transcends the status of an individual property to become a shared harmony, a cognition that arises precisely from the disparity between different worlds.

When the gray wolf and the human first began to share space, fire, and food, they could not have imagined they were initiating a cognitive co-evolution. What began as a functional alliance, an uneasy truce, transformed into the start of a long and profound shared "induction game" [Schleidt2003]. For each species, the other represented a universe of "disturbances" and contradictions. A human gesture that the wolf interpreted as a threat; a growl that the human read as aggression. Faced with this dissonance—the gap between the expected and the actual—these failures in mutual prediction activated life's most fundamental protocol in both systems: the resolution of contradiction [Putnam1966]. A random search for new responses was triggered, an exploration forced by the need to reduce uncertainty.

138

The wolf that suppressed an instinct and offered a signal of submission; the human who learned to read the movement of a tail or the tension in a pair of ears. Both, through trial and error, were generating and stabilizing new "rules" of behavior. Over time, and with repetition and variation, this cycle not only modified their conduct but, thanks to synaptic plasticity, sculpted their very neural architecture. Studies showing how dogs process certain human words in brain regions analogous to our own are the neurological fossils of thousands of years of this cognitive co-evolution [Andics2016]. The bridge was not built with bricks, but with the incessant resolution of adaptive errors, giving rise to a hybrid cognitive system that belongs neither to the dog nor to the human, but to the bond between them.

This phenomenon becomes even more extraordinary when we consider that each species inhabits its own perceptual and meaningful universe—its *Umwelt*, as biologist Jakob von Uexküll termed it [vonUexkull1957]. The world of a bat, woven from echoes and pressure waves, is fundamentally alien to our visual world. The olfactory universe of a bloodhound possesses a richness we can barely conceive. How, then, can bridges be built between such disparate realities?

Here, we must broaden our concept of the Theory of Mind (ToM) [Premack1978], seeing it not as a single milestone, but as a spectrum of "modeling the other." These bridges are built on a functional and predictive ToM. It is not that a bird or a fungus "thinks" about a tree's mental states, but that their systems have developed a highly effective predictive model of their symbiotic partner's behavior. The mycelial network that distributes nutrients to the trees in a forest does not "know" that a tree needs carbon; it simply operates under an adaptive heuristic that responds to chemical signals, a model that has proven to be evolutionarily stable [Simard2021]. These are conversations without grammar, understandings without symbols. Intelligence is not transmitted through a shared code; it emerges from the synchrony of actions and the convergence of predictive models.

But not all bridges are forged under the pressure of survival. Once the threat dissipates, the Inductive Machine does not stop; it shifts modes and activates a protocol of creative exploration: play. Play is a simulation laboratory, a safe space where rules can be broken and recombined without fatal consequences [Burghardt2005]. It is the heuristic of curiosity in its purest form, a divergent thinking in action that allows different inductive machines to explore their own limits and those of the other. The wag of a dog's tail is no longer just a stabilized rule; it is a meta-signal that communicates, "this is a simulation" [Bekoff2001]. It is the emergence of consensual fiction, the basis of all creativity.

When these bridges become stable and successful rules propagate, animal culture is born. The hunting traditions of orcas, the regional dialects among cetaceans, or the use of tools passed down through generations of chimpanzees are, in essence, collective heuristics: stabilized solutions to adaptive errors that spread socially [Whiten1999]. Active teaching, such as meerkats instructing their young on how to handle dangerous prey, is an explicit mechanism for transmitting a "patch" for a potentially fatal error [Thornton2006]. This behavior is a manifestation of antifragility at the group level: the experience of one individual who survived a stressor becomes knowledge that strengthens the entire community [Taleb2012]. Animal culture is, therefore, the collective memory of overcome errors.

Into this vast landscape of relational intelligences, Artificial Intelligence erupts as the first non-biological cognitive bridge we have consciously designed. Projects like the Cetacean Translation Initiative (CETI), which seek to decode the communication of sperm whales, are not attempting a simple word-for-word translation [CETI_ND]. Their goal is deeper: to map the semantic space of another mind, to discover its syntactic structure, and ultimately, to build a functional model of its Inductive Machine. Here, AI acts as a partner in the induction game, an artificial echo searching for patterns in a torrent of data that, to us, is just noise [Rabinowitz2018].

However, this new bridge confronts us with profound philosophical and ethical dilemmas. Even if we achieve a functional translation, can we ever access *what it is like* to be a whale—the subjective quality of its experience? Decoding does not guarantee an understanding of the *Umwelt*. Moreover, this capability imposes an unprecedented responsibility upon us. If we can listen, do we have an obligation to respond or the right to remain silent? The potential to disturb communicative ecosystems that have evolved over millennia is immense, demanding an ethics of technological cohabitation.

The journey across these cognitive bridges reveals a fundamental truth: intelligence is not a possession, but a form of attunement. It is an emergent property of the space between systems, a pattern stabilized through mutual inference, explored in play, and scaled through culture. This leaves us at a fascinating threshold. If intelligence is inherently relational, what happens when one species—ours—develops an unparalleled capacity to turn that same inferential machinery "inward"?

The emergence of symbolic language, metacognition, and the scientific method are manifestations of this introspective turn [Fleming2012]. They are the result of applying the same principles of the Inductive Machine—resolving contradictions, finding patterns, stabilizing rules—no longer to model another being, but to model our own thought and the fabric of the universe itself. As we cross this threshold, we enter the study of the cognitive heuristics that define the human experience—a new and dizzying chapter in the ceaseless story of life seeking to understand itself.

COGNITIVE HEURISTICS

The Architecture of Adaptation

Our exploration has revealed the multifaceted organization of cognition, spanning from the neural activity of a solitary bee to the symbolic structures of human language, and from a plant's defensive chemistry to the formation of animal culture. This journey has crystallized an essential insight: life doesn't merely learn; it learns how to learn.

This inherent capacity for self-reflection and adaptive meta-coding manifests as a set of implicit rules. Regardless of their complexity, living systems employ these rules to resolve contradictions, reorganize behaviors, and establish new modes of interaction with their environment. These "rules" are not rigid laws or predefined algorithms, but rather **adaptive heuristics**: practical, evolutionarily honed principles that enable organisms to function effectively in uncertain conditions. Unlike formal logical rules, heuristics do not guarantee perfect solutions, yet they significantly increase the likelihood of viable outcomes.

Furthermore, these heuristics are not isolated; they are integral components of what we've termed the **Inductive Machine**. This algorithmic model describes the adaptive operation of the cognitive system, driven by the detection and resolution of contradictions. The Inductive Machine operates through a series of functional protocols that orchestrate these heuristics.

142

The Inductive Machine Protocols

1. CONTRADICTION RESOLUTION PROTOCOL

At its most fundamental level, the Inductive Machine activates when confronted with a contradiction, serving as the primary impetus for all learning. This core response is triggered when reality diverges from expectations, manifesting as a **prediction error**, an **ambiguity**, or an **adaptive failure**. Its operation unfolds as a cascade of heuristics: initially, a surge in **motivational salience** directs attention toward the unexpected. Immediately, **automatic inhibition** suppresses the failing strategy, making way for new possibilities. Next, an **internal search** commences—a randomized exploration for alternative solutions. Finally, **synaptic plasticity** reconfigures neural circuits to embed the new, successful solution, thereby consolidating the learning.

2. HEURISTIC GENERATION PROTOCOL

When a novel solution proves successful and is repeated, the system transitions to the Heuristic Generation Protocol. This mechanism transforms a temporary fix into a stable, reusable rule. Triggered by repeated success or the need to expand a solution to new contexts, this protocol converts serendipitous outcomes into reliable strategies through useful repetition, reinforcing actions with positive results. Subsequently, contextual generalization applies the rule to new yet similar situations. The system continues to refine and adjust the heuristic via a test-and-vary mechanism, culminating in functional consolidation—the process that transforms a conscious solution into an automatic, efficient habit.

3. CREATIVE EXPLORATION PROTOCOL

143

Once a system achieves stability and possesses surplus resources, it can shift from reactive to proactive behavior, activating the Creative Exploration Protocol. This protocol serves as the engine of innovation and discovery. Activated under conditions of **prolonged stability**, **curiosity**, and **resource surplus**, it operates through **divergent thinking**, generating multiple ideas from a single starting point. It fosters **self-induced novelty** through play and the deliberate creation of new scenarios. At a more abstract level, it enables the **symbolic recombination** of concepts to forge new ones, all guided by an **anticipatory imagination** that simulates future possibilities and outcomes.

4. DIRECTED EXPLORATION PROTOCOL

However, exploration is not always unrestrained play; it often involves calculated investigation. This is the domain of the Directed Exploration Protocol, which guides an organism's cautious inquiry into novel or resource-rich environments. It is triggered by information-rich environments, contextual novelty, or endogenous motivation. This protocol directs investigation through selective curiosity, focusing attention on the most promising aspects of the new setting. The system adopts a cautious exploration—a "toe-in-the-water" approach—and performs a flexible evaluation that updates with new data. Finally, it executes an escalating response, increasing engagement only once safety or value is confirmed.

5. SOCIAL SIMULATION & COORDINATION PROTOCOL

When interaction with another being is necessary, the Social Simulation & Coordination Protocol becomes active, governing the intricate interplay between minds and forming the foundation of social life and culture. This protocol is triggered by interpersonal dissonance, the need for cooperation, or cultural transmission. It relies on the ability to model the other by constructing a predictive model of their behavior. It employs imitation as a learning accelerator and symbolic representation—language—to precisely coordinate actions. Ultimately, social memory stores collective knowledge, giving rise to culture.

6. PROACTIVE PREVENTION PROTOCOL

Learning can also be defensive. The Proactive Prevention Protocol enables systems to learn from the mistakes of others and avert danger before it materializes. Its trigger is the anticipation of risk, often learned vicariously or through a symbolic warning. It utilizes predictive memory to leverage past experiences for forecasting future threats. It employs narrative transmission—the power of stories—to instill caution. It generates anticipatory anxiety, a state of alertness that promotes prudence, and applies preventive inhibition to suppress actions deemed risky before they occur.

7. FIXATION & REDUNDANCY PROTOCOL

Finally, during periods of high volatility, the system may engage the Fixation & Redundancy Protocol. Triggered by high uncertainty or critical environments, this protocol prioritizes survival above all else. It achieves this through structural redundancy, maintaining multiple backup systems for critical functions. It uses parallel storage to distribute information and prevent total loss. It fosters mental closure, the tendency to "lock in" on a decision to reduce cognitive load during a crisis, and establishes route duplication, creating multiple pathways to achieve the same essential goal.

This collection of protocols and heuristics should be viewed not as a definitive catalog, but as an adaptable architecture, open to new combinations and expansions. Many of these processes are observable in organisms lacking a brain, such as plants, fungi, or bacteria, where the adaptive resolution of conflicts or coordination with other organisms follows analogous functional dynamics.

The significance of this synthesis lies in its explanatory power: it enables us to comprehend how life organizes its response to the unexpected, how it generates new knowledge from error, and how it extends its intelligence through collective systems. Thus, studying these heuristics not only illuminates the functional underpinnings of our own minds but also offers a shared framework for understanding all forms of life that learn.

EPILOGUE

The Mind in the Machine

In a laboratory, an echo of the digital past springs to astonishing new life. The iconic game *Pong*, a relic of the 1970s, is no longer the exclusive domain of humans or silicon algorithms. Its new player is a cluster of human brain cells, a miniature biological mind that, from a petri dish, is learning to return the ball. This image, which feels torn from a science fiction novel, is a window into a revolution in progress: Synthetic Biological Intelligence (SBI), a field that could redefine not only artificial intelligence, but our relationship with technology, biology, and consciousness itself.

The urgency for this revolution is born from an imminent crisis. Traditional computing, a master of risk management in closed systems, is running up against the immovable limits of physics. Moore's Law, the prophecy that for half a century guaranteed exponential growth, is nearing its end with circuits that measure just atoms thick. At the same time, AI's appetite for energy has become insatiable. It is estimated that by 2034, data centers will consume as much electricity as all of India. Efficiency has become the holy grail, especially when facing the uncertainty of complex, open-ended problems that traditional models cannot solve.

Facing this "silicon wall," biology emerges as a paradigm of astonishing efficiency. The human brain, with its 86 billion neurons, performs complex tasks on a mere 20 watts of power. In comparison, a supercomputer like Frontier requires 21 megawatts, making the brain approximately 500,000 times more energy-efficient. Inspired by this efficiency, labs like Cortical Labs have shown it is possible to harness it. Their creation, "DishBrain," learned to play *Pong* not through programming, but through a fundamental biological incentive, validating a key theory in neuroscience: the Free Energy Principle (FEP), a functional echo of the "endless game" that defines every inductive machine. This principle posits that living systems intrinsically act to minimize "surprise" or the unpredictability of their environment. The neurons, on their own, modified their activity to transform the chaos of a missed ball—a contradiction in the system—into the predictable order of a successful hit, thereby stabilizing a new rule of behavior.

This milestone has opened the doors to commercialization. Cortical Labs already offers its first commercial platform, the CL1, a unit that can be configured with different types of neural material, from neurons on a chip to organoids or the company's own "bio-engineered intelligence." Meanwhile, the Swiss startup FinalSpark offers remote access to brain organoids through its Neuroplatform. This new business model, dubbed "Wetware-as-a-Service" (WaaS), democratizes access to this technology, allowing researchers worldwide to experiment without the need to maintain complex laboratories.

In parallel, and perhaps more immediately, this technology is catalyzing a revolution in biomedicine. The same organoids that might one day power AI are today serving as unprecedented models for studying neurodegenerative diseases like Parkinson's and Alzheimer's, overcoming many of the limitations of animal models. They allow scientists to observe in living human tissue how diseases progress and respond to new drugs—a window that promises to accelerate the development of cures in a world where more than a third of the population will suffer from a neurological condition.

However, this promising horizon is strewn with monumental challenges. On a technical level, the lifespan of organoids is limited and their stability is fragile. Biologically, they lack internal vasculature, which leads to cell death (necrosis) at their core and impedes their growth. Furthermore, the interface with hardware is

148

an engineering challenge: current microelectrode arrays are flat (2D), making them unsuitable for effectively interacting with the complex three-dimensional structure of an organoid. The lack of standardization and reproducibility between batches is another significant barrier to large-scale adoption.

The deepest questions, however, lurk at the ethical core of the technology. As these organoids grow in complexity, could they cross the threshold into self-awareness or sentience? The problem is compounded by our difficulty in defining consciousness, an echo of the "Hard Problem" now manifested not in a brain, but in a petri dish. This creates a "regulatory quagmire" that demands legislation for an uncertain moral status, where issues like the dynamic consent of donors or the intellectual property of the data generated become critical. Moreover, biocomputing is emerging as a new "geopolitical battlefield," adding a layer of urgency to the need for global governance to prevent its misuse.

We have begun a bold journey into a future where computation will be, in part, living. The road will be long and arduous. But the *Pong* experiment has shown us that biological intelligence is far more than the sum of its parts. We are facing an era that will not only transform our machines but will force us to confront the fundamental questions of life, intelligence, and the immense responsibility of creating minds. On this journey, we have seen two cosmic forces operate in tandem: the **sieve of selection**, which with its relentless filter decides which patterns may persist, and the **sympathy of synchronization**, which weaves those persistent patterns into harmonies of ever-increasing complexity. Synthetic Biological Intelligence is merely the next step: the moment when, for the first time, we consciously become the architects of new sieves and the composers of new symphonies.

BIBLIOGRAPHIC REFERENCES

[Albert2000] Albert, D. Z. (2000). *Time and Chance.* Harvard University Press.

[Alter2017] Alter, A. (2017). *Irresistible: The Rise of Addictive Technology and the Business of Keeping Us Hooked.* Penguin Press.

[Ambrose2010] Ambrose, S. A., Bridges, M. W., DiPietro, M., Lovett, M. C., & Norman, M. K. (2010). *How Learning Works: Seven Research-Based Principles for Smart Teaching.* Jossey-Bass.

[Andics2016] Andics, A., Gábor, A., Gácsi, M., Faragó, T., Szabó, D., & Miklósi, Á. (2016). Neural mechanisms for processing lexical and intonational cues to speech in dogs. *Science, 353*(6303), 1030-1032.

[Baracchi2014] Baracchi, D., & Cini, A. (2014). The spatial organization of a honeybee colony is shaped by the brood, the queen and the season. *Journal of Insect Behavior, 27,* 587–601.

[Bassler2002] Bassler, B. L. (2002). Small talk. Cell-to-cell communication in bacteria. *Cell, 109*(4), 421-424.

[Bekoff2001] Bekoff, M., & Byers, J. A. (Eds.). (2001). *Animal Play: Evolutionary, Comparative, and Ecological Perspectives.* Cambridge University Press.

[Berridge1998] Berridge, K. C., & Robinson, T. E. (1998). What is the role of dopamine in reward: Hedonic impact, reward learning, or incentive salience? *Brain Research Reviews, 28*(3), 309–369.

[Biederman2006] Biederman, I., & Vessel, E. A. (2006). Perceptual pleasure and the brain. *American Scientist, 94*(3), 247.

[Bishop2006] Bishop, C. M. (2006). *Pattern Recognition and Machine Learning.* Springer.

[Boden2004] Boden, M. A. (2004). *The Creative Mind: Myths and Mechanisms.* Routledge.

[Borges1944] Borges, J. L. (1944). Funes el memorioso. In *Ficciones.* Editorial Sudamericana.

[Brown2009] Brown, T. (2009). *Change by Design: How Design Thinking Transforms Organizations and Inspires Innovation.* Harper Business.

[Bshary2006] Bshary, R., & Grutter, A. S. (2006). Image scoring and cooperation in a cleaner fish mutualism. *Nature, 441*(7096), 975-978.

[Bubic2010] Bubic, A., von Cramon, D. Y., & Schubotz, R. I. (2010). Prediction, cognition and the brain. *Frontiers in Human Neuroscience, 4*, art. 25.

[Buonomano2017] Buonomano, D. (2017). *Your Brain Is a Time Machine: The Neuroscience and Physics of Time*. W. W. Norton & Company.

[Burghardt2005] Burghardt, G. M. (2005). *The Genesis of Animal Play: Testing the Limits*. MIT Press.

[Buzsaki2006] Buzsáki, G. (2006). *Rhythms of the Brain*. Oxford University Press.

[Cannon1932] Cannon, W. B. (1932). *The Wisdom of the Body*. W. W. Norton & Company.

[Carroll2010] Carroll, S. (2010). *From Eternity to Here: The Quest for the Ultimate Theory of Time*. Dutton.

[Carse1986] Carse, J. P. (1986). *Finite and Infinite Games: A Vision of Life as Play and Possibility*. Free Press.

[Cashmore1999] Cashmore, A. R., Jarillo, J. A., Wu, Y. J., & Liu, D. (1999). Cryptochromes: blue light receptors for plants and animals. *Science, 284*(5415), 760-765.

[CETI_ND] CETI Project. (n.d.). *Cetacean Translation Initiative*. Retrieved from https://www.ceti.org

[Christensen1997] Christensen, C. M. (1997). *The Innovator's Dilemma: When New Technologies Cause Great Firms to Fail*. Harvard Business School Press.

[Clark2016] Clark, A. (2016). *Surfing Uncertainty: Prediction, Action, and the Embodied Mind*. Oxford University Press.

[Cowan2010] Cowan, N. (2010). The magical mystery four: How is working memory capacity limited, and why?. *Current Directions in Psychological Science, 19*(1), 51-57.

[Cremer2007] Cremer, S., Armitage, S. A., & Schmid-Hempel, P. (2007). Social immunity. *Current Biology, 17*(16), R693-R702.

[Dan2006] Dan, Y., & Poo, M. M. (2006). Spike timing-dependent plasticity: from synapse to perception. *Physiological Reviews, 86*(3), 1033-1048.

[Darwin1859] Darwin, C. (1859). *On the Origin of Species by Means of Natural Selection*. John Murray.

[Dauer2003] Dauer, W., & Przedborski, S. (2003). Parkinson's disease: mechanisms and models. *Neuron, 39*(6), 889-909.

[deWaal2016] de Waal, F. B. M. (2016). *Are We Smart Enough to Know How Smart Animals Are?*. W. W. Norton & Company.

[Deco2011] Deco, G., Jirsa, V. K., & McIntosh, A. R. (2011). Emerging concepts for the dynamical organization of resting-state activity in the brain. *Nature Reviews Neuroscience, 12*(1), 43-56.

[Dennett1991] Dennett, D. C. (1991). *Consciousness Explained*. Little, Brown and Company.

[Dill1980] Dill, L. M. (1980). The Refraction Problem in the Archerfish. *Behavioral Ecology and Sociobiology, 6*(3), 193-196.

[Droit-Volet2016] Droit-Volet, S., & Wearden, J. (2016). Passage of time judgments are not duration judgments: Evidence from a study using experience sampling methodology. *Frontiers in Psychology, 7*, art. 176.

[Dunlap1999] Dunlap, J. C. (1999). Molecular bases for circadian clocks. *Cell, 96*(2), 271-290.

[Dvornyk2003] Dvornyk, V., Vinogradova, O., & Nevo, E. (2003). Origin and evolution of circadian clock genes in prokaryotes. *Proceedings of the National Academy of Sciences, 100*(5), 2495-2500.

[Eagleman2008] Eagleman, D. M. (2008). Human time perception and its illusions. *Current Opinion in Neurobiology, 18*(2), 131-136.

[EarthSpecies_ND] Earth Species Project. (n.d.). Retrieved from https://www.earthspecies.org

[Emery2001] Emery, N. J., & Clayton, N. S. (2001). Effects of experience and social context on prospective caching by scrub jays. *Nature, 414*(6862), 443-446.

[Emery2004] Emery, N. J., & Clayton, N. S. (2004). The mentality of crows: convergent evolution of intelligence in corvids and apes. *Science, 306*(5703), 1903-1907.

[Engelberth2004] Engelberth, J., Alborn, H. T., Schmelz, E. A., & Tumlinson, J. H. (2004). Airborne signals prime plants against insect herbivore attack. *Proceedings of the National Academy of Sciences, 101*(6), 1781-1785.

[Ferguson2015] Ferguson, C. J. (2015). Does movie or video game violence predict real-life violence? It depends on your point of view. *PsycCRITIQUES, 60*(1).

[Festinger1957] Festinger, L. (1957). *A Theory of Cognitive Dissonance*. Stanford University Press.

[Feynman1965] Feynman, R. P. (1965). *The Feynman Lectures on Physics, Vol. II, Ch. 19: The Principle of Least Action*. Addison-Wesley.

[Finn2009] Finn, J. K., Tregenza, T., & Norman, M. D. (2009). Defensive tool use in a coconut-carrying octopus. *Current Biology, 19*(23), R1069-R1070.

[Fleming1929] Fleming, A. (1929). On the antibacterial action of cultures of a penicillium, with special reference to their use in the isolation of B. influenzæ. *British Journal of Experimental Pathology, 10*(3), 226-236.

[Fleming2012] Fleming, S. M., & Dolan, R. J. (2012). The neural basis of metacognitive ability. *Philosophical Transactions of the Royal Society B: Biological Sciences, 367*(1594), 1338-1349.

[Friston2010] Friston, K. (2010). The free-energy principle: a unified brain theory?. *Nature Reviews Neuroscience, 11*(2), 127-138.

[Gagliano2014] Gagliano, M., Vigoroux, V. M., Martel, A., & Col, M. (2014). Learning by association in plants. *Oecologia, 175,* 63-72.

[GarciaMarquez1967] García Márquez, G. (1967). *Cien años de soledad.* Editorial Sudamericana.

[Gardner1969] Gardner, R. A., & Gardner, B. T. (1969). Teaching sign language to a chimpanzee. *Science, 165*(3894), 664-672.

[Giacomo2019] Giacomo, C., & Volpato, G. L. (2019). Tool use in fishes. *Fish and Fisheries, 20*(3), 540-549.

[Gigerenzer2011] Gigerenzer, G., & Gaissmaier, W. (2011). Heuristic decision making. *Annual Review of Psychology, 62,* 451-482.

[Giron-Nava2023] Giron-Nava, A., et al. (2023). An abyssal octopus nursery is associated with a low-temperature hydrothermal field. *Science Advances, 9*(34), eadd8649.

[GodfreySmith2016] Godfrey-Smith, P. (2016). *Other Minds: The Octopus, the Sea, and the Deep Origins of Consciousness.* Farrar, Straus and Giroux.

[Granic2014] Granic, I., Lobel, A., & Engels, R. C. M. E. (2014). The benefits of playing video games. *American Psychologist, 69*(1), 66-78.

[Gusnard2001] Gusnard, D. A., & Raichle, M. E. (2001). Searching for a baseline: functional imaging and the resting human brain. *Nature Reviews Neuroscience, 2*(10), 685-694.

[Hamilton1964] Hamilton, W. D. (1964). The genetical evolution of social behaviour. I & II. *Journal of Theoretical Biology, 7*(1), 1-52.

[Hasselmo2006] Hasselmo, M. E. (2006). The role of acetylcholine in learning and memory. *Current Opinion in Neurobiology, 16*(6), 710-715.

[Hawking1988] Hawking, S. W. (1988). *A Brief History of Time: From the Big Bang to Black Holes.* Bantam Books.

[Hebb1949] Hebb, D. O. (1949). *The Organization of Behavior: A Neuropsychological Theory.* Wiley.

[Hiltzik1999] Hiltzik, M. (1999). *Dealers of Lightning: Xerox PARC and the Dawn of the Computer Age.* HarperCollins.

[Hjeltnes2017] Hjeltnes, B., et al. (2017). *The Health Situation in Norwegian Aquaculture 2016.* Norwegian Veterinary Institute.

[Hochner2006] Hochner, B. (2006). Octopus a model for a comparative analysis of the evolution of learning and memory mechanisms. *The Biological Bulletin, 210*(3), 308-317.

[Hohwy2013] Hohwy, J. (2013). *The Predictive Mind.* Oxford University Press.

[Holick2007] Holick, M. F. (2007). Vitamin D deficiency. *New England Journal of Medicine, 357*(3), 266-281.

[Holland1986] Holland, J. H., Holyoak, K. J., Nisbett, R. E., & Thagard, P. R. (1986). *Induction: Processes of Inference, Learning, and Discovery.* MIT Press.

[Holldobler1990] Hölldobler, B., & Wilson, E. O. (1990). *The Ants.* Harvard University Press.

[Holldobler2009] Hölldobler, B., & Wilson, E. O. (2009). *The Superorganism: The Beauty, Elegance, and Strangeness of Insect Societies.* W. W. Norton & Company.

[Holzel2011] Hölzel, B. K., Carmody, J., Vangel, M., Congleton, C., Yerramsetti, S. M., Gard, T., & Lazar, S. W. (2011). Mindfulness practice leads to increases in regional brain gray matter density. *Psychiatry Research: Neuroimaging, 191*(1), 36-43.

[Hopcroft2006] Hopcroft, J. E., Motwani, R., & Ullman, J. D. (2006). *Introduction to Automata Theory, Languages, and Computation* (3rd ed.). Pearson.

[Huang2024] Huang, T., et al. (2024). Non-neuronal memory mechanisms: CREB activation in somatic tissues. *Nature Communications, 15*, 1024.

[Hughes2011] Hughes, D. P., Andersen, S., Hywel-Jones, N. L., Himaman, W., Billen, J., & Boomsma, J. J. (2011). Behavioral mechanisms and morphological symptoms of zombie ants dying from fungal infection. *BMC Ecology, 11*(13).

[Illeris2007] Illeris, K. (2007). *How We Learn: Learning and Non-learning in School and Beyond.* Routledge.

[IntuitiveSurgicalND] Intuitive Surgical. (n.d.). *The da Vinci Surgical System.* Retrieved from https://www.intuitive.com/en-us/products-and-services/da-vinci.

[Ivry2004] Ivry, R. B., & Spencer, R. M. C. (2004). The neural representation of time. *Current Opinion in Neurobiology, 14*(2), 225-232.

[Johnson-Laird1983] Johnson-Laird, P. N. (1983). *Mental Models: Towards a Cognitive Science of Language, Inference, and Consciousness.* Harvard University Press.

[Kahneman1979] Kahneman, D., & Tversky, A. (1979). Prospect theory: An analysis of decision under risk. *Econometrica, 47*(2), 263-292.

[Kahneman2011] Kahneman, D. (2011). *Thinking, Fast and Slow.* Farrar, Straus and Giroux.

[Kandel2001] Kandel, E. R. (2001). The molecular biology of memory storage: a dialog between genes and synapses. *Science, 294*(5544), 1030-1038.

[Kandel2021] Kandel, E. R., Koester, J. D., Mack, S. H., & Siegelbaum, S. A. (2021). *Principles of Neural Science* (6th ed.). McGraw-Hill.

[Kapp2012] Kapp, K. M. (2012). *The Gamification of Learning and Instruction: Game-based Methods and Strategies for Training and Education.* Pfeiffer.

155

[Kauffman2019] Kauffman, S. A. (2019). *A World Beyond Physics: The Emergence and Evolution of Life.* Oxford University Press.

[Kessler2001] Kessler, A., & Baldwin, I. T. (2001). Defensive function of herbivore-induced plant volatile emissions in nature. *Science, 291*(5511), 2141-2144.

[Knight1921] Knight, F. H. (1921). *Risk, Uncertainty and Profit.* Houghton Mifflin.

[Koch2004] Koch, C. (2004). *The Quest for Consciousness: A Neurobiological Approach.* Roberts & Company Publishers.

[Kohda2019] Kohda, M., Hotta, T., Takeyama, T., Awata, S., Tanaka, H., Asai, J. Y., & Jordan, A. L. (2019). If a fish can pass the mark test, what are the implications for consciousness and self-awareness in animals? *PLoS Biology, 17*(2), e3000021.

[Koob2016] Koob, G. F., & Volkow, N. D. (2016). Neurobiology of addiction: a neurocircuitry analysis. *The Lancet Psychiatry, 3*(8), 760-773.

[Kosinski2023] Kosinski, M. (2023). Theory of Mind May Have Spontaneously Emerged in Large Language Models. *arXiv preprint arXiv:2302.02083.*

[Kröger2011] Kröger, B., Vinther, J., & Fuchs, D. (2011). Cephalopod origin and evolution: A congruent picture emerging from fossils, development and molecules. *Bioessays, 33*(8), 602-613.

[Krupenye2016] Krupenye, C., Kano, F., Hirata, S., Call, J., & Tomasello, M. (2016). Great apes anticipate that other individuals will act according to false beliefs. *Science, 354*(6308), 110-114.

[Krutzen2005] Krützen, M., Mann, J., Heithaus, M. R., Connor, R. C., Bejder, L., & Sherwin, W. B. (2005). Cultural transmission of tool use in bottlenose dolphins. *Proceedings of the National Academy of Sciences, 102*(25), 8939-8943.

[Kuramoto1984] Kuramoto, Y. (1984). *Chemical Oscillations, Waves, and Turbulence.* Springer-Verlag.

[Lauring2012] Lauring, A. S., Frydman, J. O., & Andino, R. (2012). The role of mutational robustness in RNA virus evolution. *Nature Reviews Microbiology, 11*(5), 327-336.

[LeDoux1996] LeDoux, J. E. (1996). *The Emotional Brain: The Mysterious Underpinnings of Emotional Life.* Simon & Schuster.

[Levin2021] Levin, M. (2021). Bioelectric signaling: Reprogrammable circuits underlying embryogenesis, regeneration, and cancer. *Cell, 184*(8), 1971-1989.

[Libet1983] Libet, B., Gleason, C. A., Wright, E. W., & Pearl, D. K. (1983). Time of conscious intention to act in relation to onset of cerebral activity (readiness-potential). The unconscious initiation of a freely voluntary act. *Brain, 106*(3), 623-642.

[Lorenz1963] Lorenz, E. N. (1963). Deterministic Nonperiodic Flow. *Journal of the Atmospheric Sciences, 20*(2), 130–141.

[Lorenz1965] Lorenz, K. (1965). *Evolution and Modification of Behavior.* University of Chicago Press.

[Mancuso2015] Mancuso, S., & Viola, A. (2015). *Brilliant Green: The Surprising History and Science of Plant Intelligence.* Island Press.

[Mancuso2019] Mancuso, S. (2019). *The Nation of Plants.* Other Press.

[Mar2011] Mar, R. A., Oatley, K., Hirsh, J., dela Paz, J., & Peterson, J. B. (2006). Bookworms versus nerds: Exposure to fiction versus non-fiction, divergent associations with social ability, and the simulation of fictional social worlds. *Journal of Research in Personality, 40*(5), 694-712.

[Margulis1970] Margulis, L. (1970). *Origin of Eukaryotic Cells.* Yale University Press.

[Mashour2013] Mashour, G. A., & Alkire, M. T. (2013). Evolution of consciousness: a contemporary, integrated perspective. *Consciousness and Cognition, 22*(4), 1215-1234.

[Mather2015] Mather, J. A., & Anderson, R. C. (2015). Play in Cephalopods. *Current Biology, 25*(1), R25-R26.

[Mäthger2009] Mäthger, L. M., Chiao, C. C., Barbosa, A., Buresch, K. C., & Hanlon, R. T. (2009). Color matching on natural substrates in cuttlefish, Sepia officinalis. *Journal of Comparative Physiology A, 195*(6), 531-545.

[Mauk2004] Mauk, M. D., & Buonomano, D. V. (2004). The neural basis of temporal processing. *Annual Review of Neuroscience, 27*, 307-340.

[McComb2001] McComb, K., Moss, C., Durant, S. M., Baker, L., & Sayialel, S. (2001). Matriarchs as repositories of social knowledge in African elephants. *Science, 292*(5516), 491-494.

[Merchant2013] Merchant, H., Harrington, D. L., & Spencer, R. M. C. (2013). Neural basis of the perception and estimation of time. *Annual Review of Neuroscience, 36*, 313-336.

[Mezulis2004] Mezulis, A. H., Abramson, L. Y., Hyde, J. S., & Hankin, B. L. (2004). Is there a universal positivity bias in attributions? A meta-analytic review of individual, developmental, and cultural differences in the self-serving attributional bias. *Psychological Bulletin, 130*(5), 711-747.

[Miller1956] Miller, G. A. (1956). The magical number seven, plus or minus two: Some limits on our capacity for processing information. *Psychological Review, 63*(2), 81-97.

[Miller1960] Miller, G. A., Galanter, E., & Pribram, K. H. (1960). *Plans and the Structure of Behavior.* Holt, Rinehart & Winston.

[Miller2001] Miller, E. K., & Cohen, J. D. (2001). An integrative theory of prefrontal cortex function. *Annual Review of Neuroscience, 24*, 167-202.

[Mirollo1990] Mirollo, R. E., & Strogatz, S. H. (1990). Synchronization of pulse-coupled biological oscillators. *SIAM Journal on Applied Mathematics, 50*(6), 1645-1662.

[Mischel1970] Mischel, W., & Ebbesen, E. B. (1970). Attention in delay of gratification. *Journal of Personality and Social Psychology, 16*(2), 329-337.

[Mischel1989] Mischel, W., Shoda, Y., & Rodriguez, M. L. (1989). Delay of gratification in children. *Science, 244*(4907), 933-938.

[Montevil2014] Montévil, M., Mossio, M., Pocheville, A., & Kauffman, S. (2016). Theoretical principles for biology: Variation. *Progress in Biophysics and Molecular Biology, 122*(1), 36-50.

[Moore2002] Moore, R. Y. (2002). Suprachiasmatic nucleus in sleep-wake regulation. *Sleep Medicine, 3*, Suppl 1, S17-S23.

[Newport2016] Newport, C., Wallis, G., Reshitnyk, Y., & Siebeck, U. E. (2016). Discrimination of human faces by archerfish (Toxotes chatareus). *Scientific Reports, 6*, 27523.

[Nolen-Hoeksema2000] Nolen-Hoeksema, S. (2000). The role of rumination in depressive disorders and anxiety. *Journal of Abnormal Psychology, 109*(3), 504-511.

[Oliveira2011] Oliveira, R. F., et al. (2011). Cichlid fishes as a model system for the study of social behavior. *Annual Review of Neuroscience, 34*, 357-379.

[Pascual-Leone2005] Pascual-Leone, A., Amedi, A., Fregni, F., & Merabet, L. B. (2005). The plastic human brain cortex. *Annual Review of Neuroscience, 28*, 377-401.

[Patel2006] Patel, A. D. (2006). Musicality, movement, and meaning in the animal world. In *Proceedings of the 9th International Conference on Music Perception and Cognition*, Bologna, Italy.

[Pathak2024] Pathak, S., & Banerjee, A. (Eds.). (2024). *Gut Microbiome and Brain Ageing*. Springer.

[Patterson1990] Patterson, F. G., & Cohn, R. H. (1990). Language acquisition by a lowland gorilla: Koko's first ten years of vocabulary development. *Word, 41*(2), 97-143.

[Pearl1988] Pearl, J. (1988). *Probabilistic Reasoning in Intelligent Systems: Networks of Plausible Inference*. Morgan Kaufmann.

[Pellis2010] Pellis, S. M., & Pellis, V. C. (2010). The Playful Brain: Venturing to the Limits of Neuroscience. *American Journal of Play, 3*(3), 329-348.

[Pepperberg2009] Pepperberg, I. M. (2009). *Alex & Me: How a Scientist and a Parrot Discovered a Hidden World of Animal Intelligence—and Formed a Deep Bond in the Process.* HarperCollins.

[PérezMercader2021] J. Pérez-Mercader, M. Becerra, M. Vázquez-Montejo, "Synthesis of protocell-like structures using visible light and synthetic chemicals," *Scientific Reports,* vol. 11, 2021.

[Peron2011] Péron, F., John, M., Sapowicz, S., Bovet, M., & Bovet, D. (2011). A study of sharing and reciprocity in Grey parrots (Psittacus erithacus). *Animal Cognition, 14*(2), 1-10.

[Piaget1952] Piaget, J. (1952). *The Origins of Intelligence in Children.* International Universities Press.

[Pittendrigh1993] Pittendrigh, C. S. (1993). Temporal organization: reflections of a Darwinian clock-watcher. *Annual Review of Physiology, 55,* 16-54.

[Plotnik2011] Plotnik, J. M., Lair, R., Suphachoksahakun, W., & de Waal, F. B. M. (2011). Elephants know when they need a helping trunk in a cooperative task. *Proceedings of the National Academy of Sciences, 108*(12), 5116-5121.

[Popper1959] Popper, K. (1959). *The Logic of Scientific Discovery.* Hutchinson.

[Premack1978] Premack, D., & Woodruff, G. (1978). Does the chimpanzee have a theory of mind?. *Behavioral and Brain Sciences, 1*(4), 515-526.

[Przybylski2017] Przybylski, A. K., Weinstein, N., & Murayama, K. (2017). Internet gaming disorder: Investigating the clinical relevance of a new phenomenon. *American Journal of Psychiatry, 174*(3), 230-236.

[Putnam1966] Putnam, P. (1966). On the Origin of Order in Behavior. *General Systems, 11,* 13–27.

[Quaranta2007] Quaranta, A., Siniscalchi, M., & Vallortigara, G. (2007). Asymmetric tail-wagging responses by dogs to different emotive stimuli. *Current Biology, 17*(6), R199-R201.

[Rabinowitz2018] Rabinowitz, N. C., Perbet, F., Song, H. F., Zhang, C., Eslami, S. M. A., & Botvinick, M. (2018). Machine theory of mind. In *Proceedings of the 35th International Conference on Machine Learning.*

[Raichle2015] Raichle, M. E. (2015). The brain's default mode network. *Annual Review of Neuroscience, 38,* 433-447.

[Reichenbach1956] Reichenbach, H. (1956). *The Direction of Time.* University of California Press.

[Rescorla1988] Rescorla, R. A. (1988). Pavlovian conditioning: It's not what you think it is. *American Psychologist, 43*(3), 151-160.

[Ries2011] Ries, E. (2011). *The Lean Startup: How Today's Entrepreneurs Use Continuous Innovation to Create Radically Successful Businesses.* Crown Business.

[Rolls2010] Rolls, E. T., & Deco, G. (2010). *The Noisy Brain: Stochastic Dynamics as a Principle of Brain Function.* Oxford University Press.

[Rovelli2018] Rovelli, C. (2018). *The Order of Time.* Riverhead Books.

[Rozin2001] Rozin, P., & Royzman, E. B. (2001). Negativity bias, negativity dominance, and contagion. *Personality and Social Psychology Review, 5*(4), 296-320.

[Rutte2007] Rutte, C., & Taborsky, M. (2007). Generalized reciprocity in rats. *PLoS Biology, 5*(7), e196.

[Ryan2000] Ryan, R. M., & Deci, E. L. (2000). Self-determination theory and the facilitation of intrinsic motivation, social development, and well-being. *American Psychologist, 55*(1), 68-78.

[Sacktor2011] Sacktor, T. C. (2011). How does PKMzeta maintain long-term memory?. *Nature Reviews Neuroscience, 12*(1), 9-15.

[Salamone2002] Salamone, J. D., & Correa, M. (2002). Motivational views of reinforcement: Implications for understanding the behavioral functions of nucleus accumbens dopamine. *Behavioural Brain Research, 137*(1-2), 3–25.

[Salas2006] Salas, C., Broglio, C., & Rodríguez, F. (2006). Neuropsychology of learning and memory in teleost fish. *Zebrafish, 3*(3), 301-314.

[Sass1992] Sass, L. A. (1992). *Madness and Modernism: Insanity in the Light of Modern Art, Literature, and Thought.* Harvard University Press.

[Schacter2007] Schacter, D. L., Addis, D. R., & Buckner, R. L. (2007). Remembering the past to imagine the future: the prospective brain. *Nature Reviews Neuroscience, 8*(9), 657-661.

[Schafer2012] Schafer, D. P., Lehrman, E. K., Kautzman, A. G., Koyama, R., Mardinly, A. R., Yamasaki, R., Ransohoff, R. M., Greenberg, M. E., Barres, B. A., & Stevens, B. (2012). Microglia sculpt postnatal neural circuits in an activity and complement-dependent manner. *Neuron, 74*(4), 691-705.

[Scheel2017] Scheel, D., Godfrey-Smith, P., & Lawrence, M. (2017). Signal use by octopuses in agonistic interactions. *Current Biology, 27*(22), 3515-3521.

[Schleidt2003] Schleidt, W. M., & Shalter, M. D. (2003). Co-evolution of humans and canids. *Evolution and Cognition, 9*(1), 57-72.

[Schultz1998] Schultz, W. (1998). Predictive reward signal of dopamine neurons. *Journal of Neurophysiology, 80*(1), 1–27.

[Seeley1982] Seeley, T. D. (1985). *Honeybee Ecology: A Study of Adaptation in Social Life.* Princeton University Press.

[Seth2021] Seth, A. (2021). *Being You: A New Science of Consciousness.* Dutton.

[Shinozuka2013] Shinozuka, K., et al. (2013). Goldfish can discriminate between music by Bach and Stravinsky. *Animal Behaviour, 86*(1), 13-19.

[Simard2021] Simard, S. W. (2021). *Finding the Mother Tree: Discovering the Wisdom of the Forest.* Knopf.

[Simon1971] Simon, H. A. (1971). Designing organizations for an information-rich world. In M. Greenberger (Ed.), *Computers, Communications, and the Public Interest* (pp. 37-72). The Johns Hopkins Press.

[Singer1999] Singer, W. (1999). Neuronal synchrony: a versatile code for the definition of relations?. *Neuron, 24*(1), 49-65, 111-125.

[Sneddon2003] Sneddon, L. U., Braithwaite, V. A., & Gentle, M. J. (2003). Do fish have nociceptors? Evidence for the evolution of a vertebrate sensory system. *Proceedings of the Royal Society B: Biological Sciences, 270*(1520), 1115-1121.

[Strogatz2003] Strogatz, S. H. (2003). *Sync: The Emerging Science of Spontaneous Order.* Hyperion.

[Strogatz2014] Strogatz, S. H. (2014). *Nonlinear Dynamics and Chaos: With Applications to Physics, Biology, Chemistry, and Engineering.* Westview Press.

[Suddendorf2007] Suddendorf, T., & Corballis, M. C. (2007). The evolution of foresight: What is mental time travel, and is it unique to humans?. *Behavioral and Brain Sciences, 30*(3), 299-313.

[Szostak2012] J. W. Szostak, "Attempts to define life do not help to understand the origin of life," J. Biomol. Struct. Dyn., vol. 29, no. 4, pp. 599–600, 2012.

[Sudhof2013] Südhof, T. C. (2013). Neurotransmitter release: the last millisecond in the life of a synaptic vesicle. *Neuron, 80*(3), 675-690.

[Taleb2007] Taleb, N. N. (2007). *The Black Swan: The Impact of the Highly Improbable.* Random House.

[Taleb2012] Taleb, N. N. (2012). *Antifragile: Things That Gain from Disorder.* Random House.

[Tedeschi2004] Tedeschi, R. G., & Calhoun, L. G. (2004). Posttraumatic growth: Conceptual foundations and empirical evidence. *Psychological Inquiry, 15*(1), 1-18.

[Terrace1979] Terrace, H. S., Petitto, L. A., Sanders, R. J., & Bever, T. G. (1979). Can an ape create a sentence?. *Science, 206*(4421), 891-902.

[Thornton2006] Thornton, A., & McAuliffe, K. (2006). Teaching in wild meerkats. *Science, 313*(5784), 227-229.

[Trewavas2005] Trewavas, A. (2005). Green plants as intelligent organisms. *Trends in Plant Science, 10*(9), 413-419.

[Uetake2001] Uetake, Y., et al. (2001). Two KaiA-binding domains of cyanobacterial circadian clock protein KaiC. *FEBS Letters, 496*(2-3), 86-90.

[Vail2013] Vail, A. L., Manica, A., & Bshary, R. (2013). Referential gestures in fish collaborative hunting. *Nature Communications, 4*, 1765.

[Vaswani2017] Vaswani, A., Shazeer, N., Parmar, N., Uszkoreit, J., Jones, L., Gomez, A. N., Kaiser, Ł., & Polosukhin, I. (2017). Attention is all you need. In *Advances in Neural Information Processing Systems 30 (NIPS 2017)*.

[Volkow2014] Volkow, N. D., & Baler, R. D. (2014). Addiction science: Uncovering neurobiological complexity. *Neuropharmacology, 76*, Part B, 235–249.

[vonBertalanffy1968] von Bertalanffy, L. (1968). *General System Theory: Foundations, Development, Applications*. George Braziller.

[vonFrisch1967] von Frisch, K. (1967). *The Dance Language and Orientation of Bees*. Harvard University Press.

[vonUexkull1957] von Uexküll, J. (1957). A Stroll Through the Worlds of Animals and Men. In C. H. Schiller (Ed. & Trans.), *Instinctive Behavior: The Development of a Modern Concept* (pp. 5-80). International Universities Press.

[Watts2018] Watts, T. W., Duncan, G. J., & Quan, H. (2018). Revisiting the Marshmallow Test: A Conceptual Replication Investigating Links Between Early Delay of Gratification and Later Outcomes. *Psychological Science, 29*(7), 1159-1177.

[Wegner2002] Wegner, D. M. (2002). *The Illusion of Conscious Will*. MIT Press.

[Weiss2016] Weiss, M. C., Sousa, F. L., Mrnjavac, N., Neukirchen, S., Roettger, M., Nelson-Sathi, S., & Martin, W. F. (2016). The physiology and habitat of the last universal common ancestor. *Nature Microbiology, 1*(9), 16116.

[Werbach2012] Werbach, K., & Hunter, D. (2012). *For the Win: How Game Thinking Can Revolutionize Your Business*. Wharton Digital Press.

[Wheeler1911] Wheeler, W. M. (1911). The ant-colony as an organism. *Journal of Morphology, 22*(2), 307-325.

[Whiten1999] Whiten, A., Goodall, J., McGrew, W. C., Nishida, T., Reynolds, V., Sugiyama, Y., Tutin, C. E. G., Wrangham, R. W., & Boesch, C. (1999). Cultures in chimpanzees. *Nature, 399*(6737), 682-685.

[Wimmer1983] Wimmer, H., & Perner, J. (1983). Beliefs about beliefs: Representation and constraining function of wrong beliefs in young children's understanding of deception. *Cognition, 13*(1), 103-128.

[Winfree1980] Winfree, A. T. (1980). *The Geometry of Biological Time*. Springer-Verlag.

[Wittmann2013] Wittmann, M. (2013). The inner sense of time: how the brain creates a representation of duration. *Nature Reviews Neuroscience, 14*(3), 217-223.

[Zhabotinsky1991] Zhabotinsky, A. M. (1991). A history of chemical oscillations and waves. *Chaos, 1*(4), 379-386.

www.ingramcontent.com/pod-product-compliance
Lightning Source LLC
Chambersburg PA
CBHW071457220526
45472CB00003B/828